Curvature Cosmology

A Model for a Static, Stable Universe

David F. Crawford

BrownWalker Press
Boca Raton • 2006

Curvature Cosmology: A Model for a Static, Stable Universe

Copyright © 2006 David F. Crawford
All rights reserved.

BrownWalker Press
Boca Raton, Florida
USA • 2006

ISBN: 1-59942-413-4 (paperback)
ISBN: 1-59942-414-2 (ebook)

BrownWalker.com

Back cover: Hubble Deep field South
Credit: R. Williams (STScI), the HDF-S Team, and NASA

Library of Congress Cataloging-in-Publication Data

Crawford, David F.
 Curvature cosmology : a model for a static, stable universe / David F. Crawford.
    p. cm.
 Includes bibliographical references and index.
  ISBN 1-59942-413-4 (pbk. : alk. paper) -- ISBN 1-59942-414-2 (ebook : alk. paper)
 1. Curvature cosmology. 2. Red shift. 3. Space and time. 4. String theory. 5. Spaces of constant curvature. I. Title.

QB991.C87C73 2006
523.1--dc22

2006036862

Contents

Contents		v
Preface		viii
1	**Introduction**	1
1.1	Other Phenomena	8
1.2	Other alternative theories	10
2	**Overview of curvature cosmology**	12
2.1	The new hypotheses	12
2.2	Curvature-redshift	12
2.3	Curvature-pressure	14
2.4	The Cosmological Model	15
3	**Derivation of curvature-redshift**	17
3.1	Introduction	17
3.2	Photons in Curved Spacetime	18
3.3	Curvature-redshift secondary photons	23
3.4	Inhibition of curvature-redshift	25
3.5	Possible laboratory tests	27
3.6	Interactions for other particles	28
3.7	Summary	30
4	**Derivation of curvature-pressure**	31
4.1	Basic hypothesis	31
4.2	Gravitation is not a force	31
4.3	A Newtonian model	32
4.4	General Relativistic model	34
4.5	Local curvature-pressure	36
5	**The curvature-cosmological model**	39
5.1	Introduction	39
5.2	The homogenous isotropic model	39
	5.2.1 The Friedmann equations	40
	5.2.2 Temperature of the universal plasma	41
	5.2.3 Hubble constant: theory	42
5.3	Geometry of curvature-cosmology	43
	5.3.1 Geometry : conclusion	45
5.4	Luminosities and magnitudes	45
5.5	Angular size	49
5.6	Surface brightness	49
5.7	Auxiliary topics	50
	5.7.1 Entropy	50

5.7.2	Olber's Paradox	52
5.7.3	Nuclear abundances	53
5.7.4	Black holes and Jets	54
5.7.5	Large number coincidences	55

6 Cosmological tests 57
- 6.1 Analytic methods — 57
  - 6.1.1 Selection effects and bias — 59
- 6.2 X–ray background — 61
- 6.3 Cosmic Microwave Radiation — 69
  - 6.3.1 CMBR temperature at large redshifts — 73
- 6.4 The Sunyaev-Zel'dovich effect — 74
- 6.5 Type 1a Supernovae — 75
  - 6.5.1 Analysis with curvature-cosmology — 78
  - 6.5.2 Supernovae: selection effects — 84
  - 6.5.3 Type 1a supernovae: magnitudes — 86
  - 6.5.4 Type 1a supernovae: SNLS data — 88
  - 6.5.5 Supernovae analysis: summary — 89
  - 6.5.6 Supernovae analysis: conclusion — 91
- 6.6 Quasar variability in time — 91
- 6.7 The linear size of radio sources — 92
- 6.8 Tolman surface brightness — 95
- 6.9 Clusters of Galaxies — 100
  - 6.9.1 Related cluster properties — 105
  - 6.9.2 Cluster of Galaxies: conclusion — 108
- 6.10 The Hubble constant — 108
- 6.11 Lyman-alpha forest — 110

7 Evolution 113
- 7.1 Evolution in curvature-cosmology — 113
- 7.2 Galaxy distributions — 114
  - 7.2.1 Observations and analysis — 114
  - 7.2.2 Galaxy evolution — 118
- 7.3 Quasar distribution — 119
  - 7.3.1 Quasar evolution — 127
- 7.4 The Butcher-Oemler effect — 129
- 7.5 Radio Source Counts — 130
- 7.6 Evolution: conclusion — 134

8 Non-cosmological tests 136
- 8.1 Galactic rotation curves — 136
- 8.2 Redshifts in our Galaxy — 137
- 8.3 Solar neutrino production — 139
- 8.4 Heating of the solar corona — 141
- 8.5 Pioneer 10 acceleration — 142

	vii
9 Conclusion	**147**
9.1 Summary of observations	147
References	**150**
Index	**158**

Preface

This book describes a new cosmological model. It is based on two major hypotheses, curvature-redshift, and curvature-pressure and the complete model is called curvature-cosmology. Also included are tests of the hypotheses that are based on non-cosmological observations.

There are no observations that refute the model. My objective is to present a new theory and thus topics are covered only to the extent that they have a bearing on the theory. For example, in many cases I have chosen what I think are the most recent and best sets of data to the exclusion of earlier results. I apologise to those authors whose work is thereby omitted.

The development of this book has taken many decades and I have received valuable input from many colleagues in the Astrophysics Department, School of Physics, University of Sydney. Unfortunately, the passage of time prevents full acknowledgement of their contributions. However, I would like to thank Dr Tony Turtle who has patiently listened to many of my hypotheses and has provided valuable criticism and encouragement. All of the analysis has been done using the Linux operating system and I am very grateful to all those programmers who have made it such a stable and useful system. The graphics have been done using the DISLIN plotting library provided by Helmut Michels at the Max-Plank-Institut in Lindau.

The advent of the internet and the use of NASA's Astrophysics Data System has greatly facilitated this work by providing access to most of the published literature in astronomy. It is equivalent to having a major library on one's desk. This research has made use of the NASA/IPAC Extragalactic Database (NED) that is operated by the Jet Propulsion Laboratory, California Institute of Technology, under contract with the National Aeronautics and Space Administration.

Finally, I wish to thank my wife June whose patience and help I greatly appreciate.

David F. Crawford : davdcraw@bigpond.net.au
Sydney, Australia
August 2006

1 Introduction

Theories of the universe date back to the dawn of civilisation. Since the advent of Einstein's theory of general relativity in 1915, there has been a major upsurge in the application of physics to cosmology. At present, the dominant cosmological model is known as the 'Big-Bang'. It is based on an unstable solution of the equations of general relativity such that the Hubble redshift is explained as due to expansion of spacetime. A major characteristic of Big-Bang theory is its flexibility, enabling it to accommodate observations with ad hoc additions, but at the same time making it a theory that is difficult to refute. Although Big-Bang cosmology has had considerable success in explaining many astronomical observations, it sometimes does so by way of additions such as inflation, dark matter, and quintessence. Such additions make it an inelegant theory. Despite this, Big-Bang cosmology is currently the dominant paradigm and nearly all analyses of cosmological observations use it either directly or indirectly.

Curvature cosmology is a complete cosmology that is presented as an alternative to the Big-Bang model. Scientifically it is a much better theory as listed below.

- It is elegant and mathematically much simpler.
- It is refutable.
- It obeys the perfect cosmological principle of being uniform in both space and time.
- It is additional to general relativity and quantum mechanics and does not require any changes to either theory.
- For homogeneous plasma, it has one free parameter – the density.
- Hubble's constant is determined by the density.
- It predicts an accurate value for the temperature of the X-ray background radiation.

- It predicts an accurate value for the temperature of the microwave background radiation.
- There is excellent agreement with many cosmological observations without the need to invoke ad hoc additional hypotheses such as evolution and dark matter.
- It can explain many minor observations such as the K-effect for bright stars.
- It accurately predicts the observed rate for the solar neutrino flux.
- It can explain the Pioneer 10 anomalous acceleration.
- A laboratory test of curvature-redshift is possible.

Apart from the initial hypotheses of curvature- redshift and curvature-pressure there are no other hypotheses needed to achieve excellent agreement with observations.

What is offered here is a cosmological theory that has an elegant structure and does not need any ad hoc additions. As a consequence, its predictions are unambiguous and the theory is potentially refutable by observational evidence. This new theory stands or falls on its own merits. Nevertheless, in testing it by comparing predictions with observations, it is often necessary to make references to Big-Bang cosmology. There are two reasons for this: one is because some observations have been carried out on the basis of demonstrating some aspect of Big-Bang cosmology, such as the search for dark matter; the second is because any new theory will necessarily need to demonstrate that the fit of the new theory to observations is superior to that of the dominant paradigm, and thus specific comparisons need to be made.

This new theory is based on two major hypotheses. The first is that the Hubble redshift is due to an interaction of photons with curved spacetime. In this sense, it is a tired-light model. The second is that there is a pressure that acts to stabilise expansion and provides a static stable universe. These two effects are called curvature-redshift and curvature-pressure. The cosmology based on these effects is hereafter referred to as curvature-cosmology. These effects do not change either general relativity or quantum mechanics: they are explicitly added to both theories.

Introduction

There are two major parts to this book. The first part describes the two hypotheses of curvature-redshift and curvature-pressure and then develops curvature-cosmology. The second part examines all the relevant cosmological data in the context of the resultant cosmological model. There is excellent agreement between the predictions and the observations. More importantly, there are no observations that refute the model. In all cases the model explains the observations as well as or much better than does the Big-Bang model.

Further applications of the hypotheses in non-cosmological contexts lead to findings that curvature-pressure can explain the solar neutrino rate deficiency. Many experiments have measured the solar neutrino flux and it is observed to be about 50% of the rate predicted by the standard solar model.

Observations of Pioneer 10 and some similar spacecraft have shown that they are consistently slowing down in a manner that is consistent with a very small acceleration towards the sun. Curvature-redshift can explain the Pioneer 10 acceleration anomaly as being due to a redshift from inter-planetary dust interacting with Doppler radar analysis.

Chapter 2 provides an overview by describing how the universe would appear in this new curvature-cosmology. The essential difference from Big-Bang cosmology is that the universe is stable, static and statistically the same at all places and at all times. It obeys the 'perfect cosmological principle'. Thus in curvature-cosmology there is no origin of the universe. In curvature-cosmology, the Hubble constant is merely a scale factor between redshift and distance and its reciprocal, which has the dimensions of time, has no relevance as an age.

The new hypothesis of curvature-redshift for photons is presented in chapter 3. The basic hypothesis is that the redshift arises from an interaction between the photons and curved spacetime. Energy is lost in each interaction to very low-energy secondary photons. A secondary hypothesis is that curvature-redshift is inhibited if there are other interactions whose mean free path is less than that for the gravitational interaction. The dominant inhibiting interaction is refractive index. The effect of curvature-redshift on other particles is also considered.

In order to observe curvature-redshift in the laboratory we need to have sufficient density of gas (or plasma) to achieve a measurable effect but not enough for there to be inhibition by the refractive index. An experiment using the Mössbauer effect for γ rays is discussed that could show the effects of curvature-redshift in a laboratory. It is shown that it is possible to measure the redshift and to detect the secondary radiation that is produced by curvature-redshift.

Chapter 4 is about curvature-pressure and uses an argument that gravitation is acceleration and not a force. Curvature-pressure has a universal application that is essential to the development of curvature-cosmology. It also has a local application that is important in explaining why the neutrino production rate in the sun is much lower than expected.

The following chapter uses the equations of General Relativity modified by the inclusion of curvature-pressure to derive the geometry of a stable static universe. It is essentially Einstein's static model with the addition of curvature-pressure, which makes it stable and finite. A very interesting consequence of the model is that it predicts that the cosmic plasma has a very high temperature of 2.56×10^9 K which agrees with observations. Another result of curvature-cosmology is that it predicts that the Hubble constant is proportional to the square root of the average density of the universe. The working equations for luminosity, angular size, and surface brightness as a function of the redshift z are derived.

The topics of entropy, Olber's Paradox, and nuclear abundances are also discussed. The model for nuclear abundances is that heavy nuclei are destroyed in the high temperature cosmic plasma to get an approximately equilibrium mix of hydrogen and helium. This collapses in sequence to clouds, clusters of galaxies, galaxies, and then stars where there is a recombination to heavier nuclei. This material is then recycled back to the cosmic plasma. The topics of black holes and astrophysical jets are briefly covered.

Chapter 6 is devoted to examining the observational evidence with two objectives in mind. The first objective is to ascertain what support there is for curvature-cosmology. However, more importantly the second objective is to see if there is any unambiguous evidence that refutes curvature-cosmology. The chapter starts with an analysis of the background X-ray observations. These are especially important in that they provide estimates of the average density and the plasma temperature. The results are in excellent agreement with curvature-cosmology.

The next topic of type 1a supernovae could be problematic in that it is generally agreed that the observations show strong evidence of time dilation due to universal expansion which is definitely contrary to curvature-cosmology. It is argued that there are strong selection effects on the observations and that, given this, the results are in excellent agreement with curvature-cosmology.

Observations show that the time scales for quasar variability are independent of redshift. This is clearly contrary to Big-Bang cosmology but in agreement with curvature-cosmology.

The distribution of the linear size of distant objects has never been satisfactorily explained in Big-Bang cosmology. However, curvature-cosmology provides a simple explanation that is in excellent agreement with observations. Next, the observed value of the Hubble constant is compared with the predicted value and again excellent agreement is found.

Because of the close relationship between luminosity and apparent area that holds in curved spacetime, observations of surface brightness provided a very important cosmological test. Currently the observations are incomplete. Nevertheless, they show strong support for curvature-cosmology.

A major argument for dark matter comes from applying the virial theorem to clusters of galaxies. Given the velocity dispersion of the galaxies, the virial theorem predicts the average mass of the galaxies. Since the predicted mass is one or two orders of magnitude larger than that predicted by their luminosities, the concept of dark matter was introduced to account for the discrepancy. In curvature-cosmology, it is shown that the discrepancy arises because the redshift dispersion has been wrongly ascribed to velocity dispersion. Curvature-redshift in the inter-galactic gas produces a redshift that is much larger than the Doppler effect of their velocities. The observed redshift dispersion is predominantly due to curvature-redshift in the inter-galactic gas. The agreement of the theory with the observations is excellent and many other velocity-related effects are also explained.

The dense spectrum of absorption lines seen in the spectra of distant quasars is ascribed to Lyman-alpha absorption in intervening clouds of gas. This is one topic where there could be serious problems for curvature-cosmology. Because of curvature-redshift within these clouds the line widths and hence the apparent absorption depths can be quite different from those derived from usual astrophysics. Hence, the full interpretation of the lines will require an analysis starting from the original spectra. What is found is that the distribution of these lines as a function of redshift does not fit the predictions. However, this distribution is extremely sensitive to the strength of the lines and since curvature-redshift cast doubts on most of the strengths this discrepancy requires a more detailed investigation.

One of the important aspects of Big-Bang cosmology is that there must be evolution, which is evidenced by change in average characteristics of objects as a function of redshift. Evolution is often invoked to explain discrepancies between observations and predictions in a rather ad hoc manner. In chapter 7 we examine the problem of evolution for galaxies in general where it is shown that with proper regard to strong selection effects there is no evidence for evolution. The next topic covered in chapter 7 is the distribution of quasar magnitudes. It is shown that in curvature-cosmology they have a well-defined distribution with a well-defined peak. This is at odds with Big-Bang cosmology where the quasar density distribution is still increasing at the faintest magnitudes. No evidence is found for evolution in the quasar data.

Many argue that one of the long-standing pieces of evidence for evolution is the Butcher-Oemler effect. This is that after corrections for band shifts, distant galaxies are bluer than nearby ones. However, recent reviews of the research literature show that the effect is not consistent and probably does not exist.

The counting of radio sources as a function of their flux density has a long history going back to the 1950s. In fact, these observations were one of the main reasons for rejecting the steady state theory of Bondi, Gold and Hoyle. Curvature-cosmology can explain all the major aspects of the distributions for a wide range of radio frequencies. This analysis provides strong evidence for the validity of the theory.

Chapter 8 covers some topics that investigate the support for curvature-redshift and curvature-pressure independent of cosmology. In Big-Bang cosmology, dark matter is used to explain the anomalous rotation curves of galaxies. In curvature-cosmology, the observations are due to a true rotation curve plus a larger curvature-redshift that is due to the halo of gas around the galaxy. The next topic is to consider the effects of curvature-redshift in our own Galaxy. For radio frequencies, the inhibition of curvature-redshift by refractive index will occur. That is, most of the radio frequency redshifts are correctly interpreted as velocities and the dynamic description of the Galaxy will be unchanged. However if there is a Galactic object that is observed at both radio and optical wavelengths the redshifts could differ. Observation of such an object that showed discrepant redshifts would provide strong support for curvature-redshift.

Even closer to home is the topic of solar neutrinos, or more particularly why is there a deficiency in their observed rates? It is shown that curvature-pressure will act to help support the solar atmosphere so that the thermodynamic pressure and hence the temperature will be less. The predicted neutrino rates are in excellent agreement with the observations. It is also shown that curvature-redshift falls short by about six orders of magnitude as an energy source for the inner solar corona. However, it may have a part to play in heating the galactic corona.

The final topic in chapter 8 is that of the anomalous acceleration of Pioneer10. In this case, there is a complex interaction between the results of curvature-redshift due to the inter-planetary dust and the navigation programs that could explain the acceleration with a plausible density of inter-planetary dust.

Chapter 9 provides a brief summary of how well the hypotheses espoused here stand up to the scrutiny of observations.

This book is the culmination of many years of work and is a complete re-synthesis of many approaches that I have already published. Because hypotheses and notations have changed and evolved, direct references to these earlier attempts to describe the theory would be misleading. Table 1 (all with author D. F. Crawford) is provided briefly stating each reference and the major topic in each paper. In nearly all cases, the data analysed in the papers has been superseded by the more recent data that are analysed in this book.

Table 1: Published papers

Year	Reference	Major topic	
1975	Nature, 254, 313	First mention of photon extent and gravity	
1979	Nature, 277, 633	Photon decay near the sun: limb effect	1
1987	Aust. J. Phys. 40, 440	First mention of curvature-redshift	2
1987	Aust. J. Phys. 40, 459	Application to background X-rays	
1991	Astrophys. J. 377, 1	More on curvature-redshift and applications	
1993	Astrophys. J. 410, 488	A static stable universe: Newtonian cosmology	
1995	Astrophys. J. 440, 466	Angular size of radio sources	
1995	Astrophys. J. 441, 488	Quasar distribution	
1999	Aust. J. Phys. 52, 753	Curvature-pressure and many other topics	

Notes:
1. Not only is the theory discredited but also the observations have not stood the test of time.
2. This gives the equation for photons but not for non-zero rest mass particles.

1.1 Other Phenomena

In Big-Bang cosmology, there are a number of basic problems and, from time to time problems posed by new data. Big-Bang cosmology is formulated in such a way that these problems can be overcome by ad hoc additions to the theory. Curvature-cosmology does not require these additions, and is formulated in such a way that good observational evidence could potentially falsify the theory. The following paragraphs briefly describe problems that occur in Big-Bang cosmology but do not occur in curvature-cosmology.

Introduction

Big-Bang cosmology has two basic problems that arise directly from the Friedmann equation. The first is the flatness problem. If the Friedmann equation is to describe a universe of the age it supposedly has, the average density must be very close to the critical density. To obtain our present universe, then at nuclear synthesis, for example, when the Universe was around 1s old, we require that the ratio of average density to critical density must be (Liddle & Lyth 2000) equal to unity to one part in $10^{16.}$ Such accuracy is unlikely to occur by chance.

The second problem is related to homogeneity of the Universe. Called the horizon problem, it arises because microwaves coming from regions separated by more than the horizon scale at last scattering, which typically subtends about a degree, cannot have interacted before decoupling (Liddle & Lyth 2000). Then why is it that the temperature of microwave radiation coming from quite different parts of the sky is so accurately the same? Guth (1981) introduced the concept of inflation as a cure for these ills. Because curvature-cosmology is stable, static and finite, it has no need for inflation.

There are three main reasons for suggested occurrence of dark matter. The first one is that since the density of observed matter is less than the critical density, there must be some unseen material that makes up the shortfall. Since critical density has no special significance in curvature-cosmology, there is no need to postulate dark matter to satisfy this shortfall. The second reason is that the quantity of matter in clusters of galaxies that is derived from using the virial theorem and the redshifts of the galaxies is much larger than the observed quantity inferred from luminosities. The third reason is to explain the anomalous rotation curves of spiral galaxies. In curvature-cosmology, both these latter observations are explained by curvature-redshift. That is the anomaly is not due to excess mass but because all of the redshift was ascribed to peculiar velocities. In a similar vein, dark energy has been suggested as a cause for the apparent acceleration of the rate of expansion of the universe. As discussed later not only is the expansion zero, there is no evidence for 'acceleration' so that dark energy (or anything else such as quintessence) is not needed.

Black holes have been a very interesting theoretical construct for many decades and many claim that there is considerable observational evidence for their existence. However, all of this evidence is indirect with the most compelling evidence being X-rays and other radiation coming from accretion disks. The evidence for the mass of black holes comes from redshifts from nearby stars and gas. However if most of these redshifts are due to curvature-redshift in the surrounding gas, these mass estimates would be considerably decreased. One consequence of curvature-pressure is that it will prevent a compact object denser than a neutron star from completely collapsing to a black hole. Instead, there would be a very dense object with a size between a neutron star and a black hole. This object would be surrounded by an accretion disk. Thus, observations interpreted as a being due to a black hole could be observations of such an object.

1.2 Other alternative theories

After reading a draft version of this book Professor Jean-Claude Pecker suggested that it would be enhanced by a comparison with other alternative theories that try to explain the Hubble redshift and other cosmological observations. Much of the early work has been reviewed by Ellis (1984). More recently Peebles (1993) provides an even handed summary. Ghosh (1991) has a list of non-velocity redshift mechanisms that have been suggested from 1919 to 1986. The majority of these alternative cosmological models are of little relevance to curvature-cosmology since they require a different or modified gravitational theory, whereas curvature-cosmology is strongly based on general relativity.

Alternative redshift theories that are or may be relevant to curvature-cosmology can be classified as tired-light theories. These have the major characteristic that as photons propagate they lose energy by some form of interaction with another particle such that the energy loss rate is proportional to the photon energy. The essential difference between these theories is in defining the type and properties of such particles. The major objection to these theories has been that the interaction will also produce an angular scattering of the photons (Zel'dovich 1963, Schatzman 1970) which is not observed.

Pecker & Vigier (1987) propose an interaction between a massive photon and vacuum particles that avoids the scattering problem. LaViolette (1986) starts from a field theory which he calls subquantum kinetics that predicts that photons travelling through intergalactic space, where the gravitational field potential is least negative, should gradually decrease their energy. Since the magnitude of the scattering is inversely proportional to the mass of the interacting particle he escapes the scattering problem by, in effect, having the photons interact with the gravitational field that has no recoil momentum so that there is no scattering. This is similar to the way in which curvature-redshift avoids the scattering problem.

A different approach has been taken by Marmet (1988, see also Marmet & Reber, 1989). He argues that the energy loss is via very low energy photons lost by bremsstrahlung during normal electromagnetic (Compton) interactions. He further argues, with support from Jauch & Rohrlich (1980), that even in 'elastic' scattering there is production of these very low energy photons. His estimate of the density of gas required to produce the Hubble redshift is about 2.5×10^4 hydrogen atoms m^{-3}, which is about 10^4 times the density, found by curvature-cosmology. Finally, he argues that the angular scattering is small enough to pass the scattering test.

Ghosh (1991) has considered a process that is based on Sciama's model for inertial induction (see section 5.7.5). Sciama added an acceleration dependent term to the Newtonian gravitational equation in his investigations of the origin of inertia. Ghosh extends this by adding a velocity dependent term. The result is a cosmic drag force that could explain the Hubble redshift. Since it is a simple drag, there is no scattering and thus no blurring of distant objects. He calls his theory velocity-dependent inertial induction and also applies other redshifts and planetary phenomena. Overall it is an interesting extension of Sciama's ideas.

Many other mechanisms such as electromagnetic and plasma interactions and photon decay have been proposed but it is difficult to find a model that provides an energy-loss rate that is proportional to the photon energy and also solves the scattering problem. By comparison, it will be shown that curvature-redshift has better agreement with current observations and it will occur in many more situations where redshifts are observed.

2 Overview of curvature cosmology

2.1 The new hypotheses

This chapter provides a qualitative description of curvature-cosmology and describes how it leads to a model of the universe that is static and stable. Thus, the major difference from Big-Bang cosmology is that there was no beginning to the universe and there is no evolution of the universe and definitely no expansion. Although there is evolution of stars and galaxies, the statistical properties of the universe are the same at all places and at all times. Hence, curvature-cosmology obeys the 'perfect cosmological principle'.

2.2 Curvature-redshift

The first hypothesis is that the redshift in the spectral lines that is observed in distant galaxies and quasars is due to a gravitational interaction. In this respect is a tired-light theory. The essence of the hypothesis is that the passage of a photon through curved space-time can be described as being the same as that for a bundle of geodesics. Then, if space-time is curved, the cross sectional area of this bundle of geodesics will change along the trajectory. For positive mass and energy, the change will be a decrease in area, a phenomenon that is described by the 'focussing theorem' (Misner, Thorne, & Wheeler, 1973). The curvature-redshift hypothesis is that characteristics of the photon, in particular the angular momentum, are computed by volume integrals and, if the scale size (as determined by the geodesic bundle) changes, then these characteristics must change. However, quantum mechanics requires that the value of the angular momentum be fixed. This leads to an impasse that is resolved by an interaction between the photon and curved spacetime. What happens is that from time to time, as determined by the uncertainty principle, the photon splits into one photon with most of the energy and two very low-energy photons. It is the loss of energy to these secondary low-energy photons that produces the observed redshifts. Although not strictly correct, it is convenient to consider the photon with most of the energy as being the continuation of the original photon but with a slightly reduced energy. Note that there must be at least two secondaries in order to conserve spin. Furthermore, on grounds of symmetry, it is hypothesised that the two secondaries are emitted in opposite directions so that there is no deviation in the 'primary' photon's trajectory.

One of the major objections to tired-light explanations (Zel'dovich 1963) for the redshift has been that if the energy loss were due to an interaction of a photon with a particle there would be angular scattering of the original photon. A typical type of interaction would be Compton scattering of the photons off charged particles. The observations of point images of very distant quasars show that this angular scattering is not observed. For curvature-redshift, there are two reasons why such scattering is expected to be negligible.

The first is that the interaction is between the photon and curved spacetime, the effective mass of which is very much larger than that of any atomic particle, and since the size of the angular scattering is inversely proportional to the scattering particle's mass the angular scattering is very small. The second reason is that it is hypothesised that the emission of the secondaries is symmetric, with zero transverse momentum that results in no deflection of the 'primary' photon.

It is shown below that for most astrophysical applications the energy loss rate and hence the redshift is proportional to the square root of the density of the medium. This result follows directly from using the 'focussing theorem' and General Relativity. This redshift mechanism is called curvature-redshift in order to distinguish this redshift from other redshifts such as those due to velocity and gravitation. The first test of General Relativity was the observation of the angular deflection of photons whose trajectory passed near the limb of the sun. Obviously, we might ask whether this deflection is accompanied by a decrease in energy. The answer is that under the hypothesis of curvature-redshift there is no predicted energy loss that would be observed as a redshift. What happens is that the cross sectional area of the geodesic bundle has its shape distorted but there is no change in its area and therefore no predicted curvature-redshift. There is however the prediction of a very small redshift due to the density of plasma in the solar corona.

Since quantum mechanics shows that particles, such as electrons and protons, can be considered as waves, curvature-redshift will apply to them in a similar manner. The only difference from photons is that the change in cross sectional area uses the geodesic equations appropriate for massive particles.

2.3 Curvature-pressure

If the redshifts of quasars and distant galaxies are due to curvature-redshift and not due to universal expansion then it is reasonable to consider static cosmological models. Although we could consider an expanding universe that includes curvature-redshift, this complication should be considered only if demanded by observations. The Einstein static cosmology, which has the same geometry as the expanding model but without expansion, has a serious defect. It is intrinsically unstable. Any slight perturbation will cause the universe to either expand or contract. Thus, any viable cosmological model that is static must address this issue. My second major hypothesis does this by introducing curvature-pressure. This not only solves the stability problem but leads to a cosmological model that is in excellent agreement with observations.

Consider the pressure at the bottom of a column of water in the ocean. Usually we ascribe the pressure as being due to the gravitational attraction of the earth acting on the water column. Now assume that the bottom is removed and the identical column of water is in free fall. There is no change whatsoever in the gravitational attraction, but the pressure has gone. This shows that the pressure is actually due to elastic forces in the bottom counteracting the acceleration due to gravity. These elastic forces are distributed throughout the earth so that the whole earth experiences the reaction force to the weight of the column of water. This thought experiment suggests that whenever a non-gravitational force opposes a gravitational acceleration there is the possibility of a pressure. For Newtonian gravity whenever some material is not in free fall there must be a pressure (or force) present. In General Relativity the concept of free fall is the same as that of geodesic motion. That is a particle with negligible mass is in 'free fall' if it follows a geodesic.

Now consider a very large volume of gas with uniform density, which determines the local curvature of spacetime. Except for particle-particle collisions, the gas particles follow geodesics and the collisions produce the usual thermodynamic pressure. There are no other forces and hence no additional pressure. However, suppose that the gas is partially or fully ionised plasma. Since the electro-magnetic forces between the particles produce accelerations that are very many orders of magnitude stronger than gravitational accelerations, the charged particles do not travel along geodesics. The hypothesis of curvature-pressure is that since the particles are no longer travelling along geodesics there is a pressure generated by the non-geodesic motion. In this case, the reaction to these non-geodesic accelerations is a curvature-pressure that is acting on the matter producing curved spacetime in such a way as to try to decrease the curvature. The pressure must act in this way or else there would be a runaway effect and violation of conservation of energy. In other words, the plasma produces curved spacetime through its density entering the stress-energy tensor in Einstein's field equations and the magnitude of the curvature increases with increasing density. Then the failure of the particles in the plasma to follow geodesics produces a pressure that acts to decrease the curvature and hence decrease the density of the plasma. In Newtonian terminology, the curvature-pressure opposes the mutual gravitational attraction of the plasma particles. It must be emphasised that although curvature-pressure acts and has many similar properties to thermodynamic pressure it is quite distinct. For example, there is no curvature-pressure in a neutral gas.

2.4 The Cosmological Model

The simplest cosmological model is a universe containing plasma with uniform density. The curvature-cosmology model is Einstein's static solution to the equations of General Relativity with the addition of curvature-pressure. This solution is described by the Friedmann equations with an additional term that describes the effects of curvature-pressure. It is the curvature-pressure that stabilises the static solution.

One of the remarkable results of curvature-cosmology is that it has one free parameter, which can be taken to be the average density of the universe. It is convenient to express this density as the equivalent number of hydrogen atoms per cubic metre. That is, define $N=\rho/m_H$, where ρ is the density and m_H is the mass of a hydrogen atom. Not only does the density determine the size of the universe it also determines the Hubble constant. For example with, $N=2$ m^{-3} we find that it predicts the Hubble constant to be $H=59$ km.s^{-1}.Mpc^{-1}. Furthermore, curvature-cosmology predicts that the plasma has a very high temperature (2.56×10^9 K). It is interesting to note that this temperature is independent of the density of the universe.

The basic cosmological model is one in which the cosmic plasma dominates the mass distribution and hence the curvature of spacetime. In this first order model, the gravitational effects of galaxies are neglected. The geometry of this curvature-cosmology is that of a three-dimensional surface of a four-dimensional hypersphere. It is almost identical to that for Einstein's static universe. For a static universe, there is no ambiguity in the definition of distances and times. One can use a universal cosmic time and define distances in light travel times or any other convenient measure. Furthermore, curvature-pressure makes the solution stable.

Curvature-cosmology makes quite specific predictions that can be refuted. Curvature-cosmology obeys the perfect cosmological principle of being homogeneous in space and constant in time. Thus, any observations that unambiguously show changes in the universe with redshift would invalidate curvature-cosmology.

In curvature-cosmology, there is a continuous process in which some of the cosmic plasma will aggregate to form galaxies and then stars. The galaxies and stars will evolve and eventually all their material will be returned to the cosmic plasma. Thus, a characteristic of curvature-cosmology is that although individual galaxies will be born, live and die, the overall population will be statistically the same for any observable characteristic.

3 Derivation of curvature-redshift

3.1 Introduction

The derivation of curvature-redshift is based on the fundamental hypothesis of Einstein's General Theory of Relativity that spacetime is curved. As a consequence, the trajectories of initially parallel point particles, geodesics, will move closer to each other as time increases. Consequently the cross sectional area of a bundle of geodesics will slowly decrease. In applying this idea to photons, we assume that a photon is described in quantum mechanics as a localised wave where the geodesics correspond to the rays of the wave. Note that this wave is quite separate from an electromagnetic wave that corresponds to the effects of many photons. It is fundamental to the hypothesis that we can consider the motion in spacetime of individual photons. Because the curvature of spacetime causes the focussing of a bundle of geodesics, this focussing also applies to a wave. As the photon progresses, the cross sectional area of the wave associated with it will decrease. However, in quantum mechanics properties such as angular momentum are computed by an integration of a radial coordinate over the volume of the wave. If the cross sectional area of the wave decreases, then the angular momentum will also decrease. However, angular momentum is a quantised parameter that has a fixed value. The solution to this dilemma is for the photon to split into two very low-energy photons and a third that has the same direction as the original photon. It is convenient to consider the interaction as a primary photon losing a small amount of energy into two secondary photons. Averaged over many photons this energy loss will be perceived as a small decrease in frequency.

Since in quantum mechanics electrons and other particles are considered as waves, a similar process will also apply. It is argued that electrons will interact with curved spacetime to lose energy by the emission of very low-energy photons.

3.2 Photons in Curved Spacetime

Einstein's General Theory of Relativity requires that the metric of spacetime be determined by the distribution of mass (and energy). In general this spacetime will be curved such that in a space of positive curvature nearby geodesics that are initially parallel will come closer together as the reference position moves along them. This is directly analogous to the fact that on the earth lines of longitude come closer together as they go from the equator to either pole. In flat spacetime, the separation remains constant. For simplicity, let us consider geodesics in a plane. Then the 'equation for geodesics deviation' can be written (Misner, Thorne, & Wheeler, 1973, p 30) as

$$\frac{d^2\xi}{ds^2} = -\frac{\xi}{a^2} \quad (1)$$

where ξ is normal to the trajectory and s is measured along the trajectory. The quantity $1/a^2$ is the Gaussian curvature at the point of consideration. For a surface with constant curvature, that is the surface of a sphere, the equation is easily integrated to get (ignoring a linear term)

$$\xi = \xi_0 \cos(s/a) \quad (2)$$

Note that this equation also describes the separation of lines of longitude as we move from the equator to either pole. Now geodesics describe the trajectories of point particles. Null-geodesics are associated with mass-less particles. However, photons are not point particles. The experiment of using single photons in a two-slit interferometer shows that individual photons must have a finite size.

Quantum mechanics requires that all particles are described by wave functions and therefore we must consider the propagation of a wave in spacetime. Because photons are bosons, the usual quantum mechanical approach is to describe the properties of photons by creation and destruction operators. The emphasis of this approach is on the production and absorption of photons with little regard to their properties as free particles. Indeed because photons travel at the speed of light, their lifetime in their own reference frame between creation and destruction is zero. However, in any other reference frame they behave like normal particles with definite trajectories and lifetimes. Havas (1966) has

Photons in Curved Spacetime

pointed out that the concept of a single photon is rather tenuous. There is no way we can tell the difference between a single photon and a bundle of photons with the same energy, momentum, and spin. However, it is an essential part of this derivation that a single photon has an actual existence.

Assume that a photon can be described by a localised wave packet that has finite extent both along and normal to its trajectory. This economic description is sufficient in the following derivation. We define the frequency of a photon as $v = E/h$ and its wavelength as $\lambda = hc/E$ where E is its energy. These definitions are for convenience and do not imply that we can ascribe a frequency or a wavelength to an individual photon; they are properties of groups of photons. The derivation requires that the wavelength is short compared to the size of the wave packet and that this is short compared to variations in the curvature of spacetime. Furthermore, we assume that the rays of any wave follow null geodesics and therefore any deviations from flat spacetime produce change in shape of the wave packet. In other words, since the scale length of deviations from flat space are large compared to the size of the wave packet they act as a very small perturbation to the propagation of the wave packet.

Consider a wave packet moving through a spacetime of constant positive curvature. Because of geodesic deviation, the rays come closer together as the wave packet moves forward. They are focussed. In particular the direction θ, of a ray (geodesic) with initial separation ξ_0 after a distance s is (assuming small angles)

$$\theta = -\frac{s\xi_0}{a^2} \qquad (3)$$

where a is the local radius of curvature. Since the geodesic is the direction of energy flow, we can integrate the wave-energy-function times the component of θ normal to the trajectory, over the dimensions of the wave packet in order to calculate the amount of energy that is now travelling normal to the trajectory. The result is a finite energy that depends on the average lateral extension of the wave packet, the local radius of curvature, and the original photon energy. The actual value is not important but rather the fact that there is a finite fraction of the energy that is moving away from the trajectory of the original wave packet. This suggests a photon interaction in which the photon interacts with curved spacetime with the hypothesis that the energy flow normal to the trajectory goes into the emission of secondary photons normal to its trajectory. From a quantum-mechanical point of view, there is a strong argument that some interaction must take place. If the spin of the photon is directly related to the angular momentum of the wave packet about its trajectory then the computation of the angular momentum is a similar integral. Then because of 'focussing' the angular momentum clearly changes along the trajectory, which disagrees with the quantum requirement that the angular momentum, that is the spin, of the photon is constant. The Heisenberg uncertainty principle requires that an incorrect value of spin can only be tolerated for a finite time before something happens to restore the correct value. We now consider the consequences.

Consider motion on the surface of a three dimensional sphere. As described above, two adjacent geodesics will move closer together due to focussing. Simple kinematics tells us that a body with velocity v associated with these geodesics has acceleration v^2/r, where r is the radius of curvature. This acceleration is directly experienced by the body. In addition, it experiences a tidal acceleration within itself. This tidal acceleration is equivalent to the focussing of the geodesics. Although the focussing and acceleration are closely linked, we need to consider whether the occurrence of one implies the occurrence of the other. Does the observation of focussing (tidal acceleration) imply acceleration in the orthogonal direction? It is true in two and three dimensions, but it needs to be demonstrated for four dimensions.

The geometry of a three dimensional surface with curvature in the fourth dimension is essentially the same as motion in three dimensions except that the focussing now applies to the cross-sectional area and not to the separation. Does this acceleration have the same physical significance? Assuming it does, a wave packet that is subject to focussing has acceleration in an orthogonal dimension. For instance if we could constrain a wave packet (with velocity c) to travel on the surface of a sphere in three dimensions it would not only show a focussing effect but also experience an acceleration of c^2/r normal to the surface of the sphere. Then a wave packet (and hence a photon) that has its cross-sectional area focussed by curvature in the fourth dimension with radius r would have an energy loss rate proportional to this acceleration. The essence of curvature-redshift hypothesis is that the tidal distortion causes the photon to interact and that the energy loss rate is proportional to c^2/r. For a photon with energy E the loss rate per unit time is cE/r, and per unit distance it is R/r.

In General Relativity the crucial equation for the focussing of a bundle of geodesics was derived by Raychaudhuri (1955, also see Misner, Thorne, & Wheeler 1973, Ellis, 1984) and for the current context we can assume that the bundle has zero shear and zero vorticity. Since any change in geodesic deviation along the trajectory will not alter the direction of the geodesics we need only consider the cross-sectional area (A) of the geodesic bundle to get the equation

$$\frac{1}{A}\frac{d^2 A}{ds^2} = -R_{\alpha\beta}U^\alpha U^\beta = -\frac{1}{a^2} \qquad (4)$$

where $R_{\alpha\beta}$ is the Ricci tensor (it is the contraction of the Riemann-Christoffel tensor), U_α is the 4-velocity of the reference geodesic and a is the local radius of curvature. This focussing can be interpreted as the second order rate of change of cross-sectional area of a geodesic bundle that is on the three-dimensional surface in four-dimensional space. Then if we consider that a photon is a wave packet we find that the rate at which the photon loses energy per unit distance is E/a or more explicitly

$$\frac{1}{E}\frac{dE}{ds} = -\frac{1}{a} = -\left(R_{\alpha\beta}U^\alpha U^\beta\right)^{1/2}, \qquad (5)$$

What is interesting about this equation is that, for the Schwarzschild (and Kerr) solutions for the external field for a mass, the Ricci tensor is zero; hence, there is no focussing and no energy loss. A geodesic bundle passing a mass such as the sun experiences a distortion but the wave packet has not changed in area. Hence, this model predicts that photons passing near the limb of the sun will not suffer any energy loss due to the gravitational effects of the sun.

The field equation for Einstein's general theory of gravitation is

$$R_{\alpha\beta} = 8\pi G\left(T_{\alpha\beta} - \frac{1}{2}T g_{\alpha\beta}\right) + \Lambda g_{\alpha\beta}, \qquad (6)$$

where T is the contracted form of $T_{\alpha\beta}$ the stress-energy-momentum tensor, $g_{\alpha\beta}$ is the metric tensor, G is the Newtonian gravitational constant and Λ is the cosmological constant. It states that the Ricci tensor describing the curvature of spacetime is determined by the distribution of mass (and energy). Direct application of the field equations (without the cosmological constant) in terms of the stress-energy-momentum tensor $T_{\alpha\beta}$, the metric tensor $g_{\alpha\beta}$ and with the material having a 4-velocity V_α gives

$$\frac{1}{a^2} = 8\pi G\left(T_{\alpha\beta}U^\alpha U^\beta - \frac{1}{2}T g_{\alpha\beta}V^\alpha V^\beta\right). \qquad (7)$$

For null geodesics $g_{\alpha\beta}V_\alpha V_\beta$ is zero which leaves only the first term. For a perfect fluid the stress-energy-momentum tensor is

$$T_{\alpha\beta} = \frac{p}{c^2}g_{\alpha\beta} + \left(\rho + \frac{p}{c^2}\right)U_\alpha U_\beta, \qquad (8)$$

where p is the proper pressure and ρ is the density. Combining equation (7) with equation (8) gives for null geodesics

$$\frac{1}{a^2} = \frac{8\pi G}{c^2}\left(\rho + \frac{p}{c^2}\right). \qquad (9)$$

For cases where the proper pressure is negligible compared to the density we can ignore the pressure and get

$$\frac{1}{E}\frac{dE}{ds} = -\frac{1}{a} = -\left(\frac{8\pi G\rho}{c^2}\right)^{1/2} = -1.366\times 10^{-13}\sqrt{\rho}\ \text{m}^{-1}. \qquad (10)$$

For many astrophysical types of plasma, it is useful to measure density by the equivalent number of hydrogen atoms per cubic metre: that is we can put $\rho = Nm_H$ and get

$$\boxed{\frac{1}{E}\frac{dE}{ds} = -\left(\frac{8\pi G N m_H}{c^2}\right)^{1/2} = -5.588\times 10^{-27}\sqrt{N}\ \text{m}^{-1}} \qquad (11)$$

where the box shows that this is one of the basic equations in curvature-cosmology. The rate of energy loss depends only on the square root of the density of the material, which may consist of gas, plasma, or gas and dust.

3.3 Curvature-redshift secondary photons

The above derivation does not define the form of energy loss. The most realistic model is that the photon decays into three secondary photons, one of which takes nearly all the energy and momentum and two very low-energy secondary photons. It is convenient (although not strictly correct) to think of the high-energy secondary as a continuation of the primary but with slightly reduced energy. Two secondary photons are required to preserve spin and, by symmetry, they are emitted in opposite directions with the same energy[1] and at right angles to the trajectory. Thus, the 'primary' photon is not deflected. We can get an estimate of how often these interactions occur and hence what the secondary energies are by using the Heisenberg uncertainty principle applied to the primary. For linear momentum and distance, it is $\Delta p \Delta x \cong h/4\pi$ and putting, $X = \Delta x$ we get $\Delta E = hc/4\pi X$. Now after the photon with energy E_0 has travelled a distance X the energy-loss is $\Delta E = E_0 X/a$ hence we get

$$X^2 = \frac{ahc}{4\pi E_0} = \frac{a\lambda_0}{4\pi} = \frac{c\lambda_0}{4\pi\sqrt{8\pi G\rho}}. \qquad (12)$$

If each secondary photon takes half the energy-loss, we find

$$\Delta E = \frac{1}{2}\frac{E_0 X}{a}. \qquad (13)$$

Therefore the secondary photons have a wavelength of

$$\lambda = \frac{2\lambda_0 a}{X} = 8\pi X = 4\sqrt{\pi a \lambda_0}. \qquad (14)$$

For example consider a visible photon with wavelength 600 nm travelling in plasma with density N, then $X = 2.92 \times 10^9 N^{-1/2}$ m and the wavelength is $\lambda = 7.34 \times 10^{10} N^{-1/2}$ m which corresponds to

---

[1] This assumption that the two secondary photons have the same energy is made without proper justification. What can be said is that if they are not they will still have nearly equal energies because the probability of having one with a much longer relative wavelength is very low.

a frequency of $v = 4.08 N^{1/2}$ mHz. Now for fully ionised plasma the plasma frequency is

$$v_p = \left(\frac{Ne^2}{\pi m_e}\right)^{1/2} = 8.975 N^{1/2} \text{ Hz}, \quad (15)$$

and the ratio is

$$\frac{v}{v_p} = 4.55 \times 10^{-4}. \quad (16)$$

Thus, in this case the secondary photons have frequencies below the plasma frequency and therefore cannot propagate but will be quickly absorbed by the plasma. The energy lost by the primary photon is dissipated into heating the plasma.

We can extend this simple analysis by noting that the number of modes of oscillation for standing electromagnetic waves in a large box is proportional to the square of the frequency. Now, as the primary photon propagates the amount of energy available for the production of secondary photons increases at a rate proportional to the distance. If we assume that all modes are equally likely, the probability of secondary emission will increase as the square of the energy available and hence it increases as the square of the distance travelled. Therefore the probability of emission is proportional to the probability that the distance x has been covered without any emission before it has reached x. Since the energy available is proportional to the distance, this can be expressed as

$$p(x)dx \propto \left(1 - \int_0^x p(x)dx\right) x^2. \quad (17)$$

This equation can be solved to get

$$p(x)dx = \frac{1}{x_1}\left(\frac{x}{x_1}\right) \exp\left(-\left(\frac{x}{x_1}\right)^3\right) dx, \quad (18)$$

where x_1 is a constant to be determined. Since we are assuming that both secondary photons have the same energy, the secondary photon frequency distribution has the same form, namely

$$p(v)dv = \frac{1}{v_1}\left(\frac{v}{v_1}\right) \exp\left(-\left(\frac{v}{v_1}\right)^3\right) dv, \quad (19)$$

where v_1 is a constant to be determined. Equation (19) can be integrated to get the average of powers of the frequency, namely

$$\langle v^n \rangle = \Gamma\left(\frac{n+3}{3}\right) v_1^n. \tag{20}$$

Hence, the average frequency is $\langle v \rangle = \Gamma(4/3) v_1$ where Γ is the gamma function. The next step is to determine the constant v_1 (and implicitly x_1). Now equation (12) provides the average distance travelled between interactions and therefore the average energy loss ΔE is computed from equation (13) giving

$$v_1 = \frac{1}{4\Gamma(4/3)} \sqrt{\frac{cv_0}{\pi a}}. \tag{21}$$

For an accurate derivation, a full treatment is required. Nevertheless, the above derivation results in a plausible indication of the form of the true spectrum.

3.4 Inhibition of curvature-redshift

From the discussion above it is clear that the process of curvature-redshift requires a gradual focussing to a critical limit, followed by the emission of secondary photons. It is as if the photon gets slowly excited by the focussing until the probability of secondary emission becomes large enough for it to occur. If there is any other interaction the excitation due to focussing will be nullified That is, roughly speaking, curvature-redshift interaction requires an undisturbed path length of at least X (equation (12)) for significant energy loss to occur. A suitable criterion for inhibition to occur is that the competing interaction has an interaction length less than X. Although Compton or Thompson scattering are possible inhibitors there is another interaction that has a much larger cross-section. This is the coherent multiple scattering that produces refractive index.

In classical electro-magnetic theory, the refractive index of a medium is the ratio of the velocity of light in vacuum to the group velocity in the medium. However, in quantum mechanics photons always travel at the velocity of light in vacuum. In a medium, a group of photons appears to have a slower velocity because the individual photons interact with the electrons in the medium and each interaction produces a time delay. Because the interaction is with many electrons spread over a finite volume, the only possible result of each interaction is the emission of another photon with the same energy and momentum. Now consider the absorption of a wave. In order to cancel the incoming wave a new wave with the same frequency and amplitude but with opposite phase must be produced. Thus, the outgoing wave will be delayed by half a period with respect to the incoming wave. For example if the phase difference was not exactly half a period for an electromagnetic wave incident on many electrons, the principle of conservation of energy would be violated. This simple observation enables us to compute the interaction length for refractive index n. If L is this interaction length then it is

$$L = \frac{\lambda_0}{2|n-1|}, \quad (22)$$

where n is the refractive index and the modulus allows for plasma and other materials where the refractive index is less than unity. Note that L is closely related to the extinction length derived by Ewald and Oseen (Jackson 1975; Born & Wolf 1999) which is a measure of the distance needed for an incident electromagnetic wave with velocity c to be replaced by a new wave. For plasmas the refractive index is

$$n \cong 1 - \frac{N_e \lambda_0^2}{2\pi r_0}, \quad (23)$$

where N_e is the electron density and r_0 is the classical electron radius. We can combine these two equations to get

$$L = (N_e r_0 \lambda_0)^{-1}. \quad (24)$$

Thus, we would expect the energy loss to be inhibited if the average curvature-redshift interaction distance is greater than that for refractive index interactions, i.e. if $X>L$. Therefore, we can compute the ratio (assuming a plasma with $N \approx N_e$) using equation (12) to get

$$X/L = 0.0106 N^{3/4} \lambda_0^{3/2}. \quad (25)$$

This result shows that curvature-redshift will be inhibited if this ratio is greater than one, which is equivalent to $\lambda_0 \geq 20.7 N^{-1/2}$ m. For example, curvature-redshift for the 21 cm hydrogen line will be inhibited if the electron density is greater than about 10^4 m^{-3}.

3.5 Possible laboratory tests

It is apparent from the above analysis that to observe the redshift in the laboratory we need to have sufficient density of gas (or plasma) to achieve a measurable effect but not enough for there to be inhibition by the refractive index. The obvious experiment is to use the Mössbauer effect for γ rays that enables very precise measurement of their frequency. Simply put, the γ rays are emitted by nuclei in solids where there is minimal recoil or thermal broadening of the emitted γ ray. Since the recoil-momentum of the nucleus is large compared to the atomic thermal energies and since the nucleus is locked into the solid so that the recoil momentum is precisely defined, then the γ ray energy is also precisely defined. The absorption process is similar and has a very narrow line width. Such an experiment has already been done by Pound & Snyder (1965). They measured gravitational effects on 14.4 keV γ rays from ^{57}Fe being sent up and down a vertical path of 22.5 m in helium near room pressure. They found agreement to about 1% with the predicted fractional redshift of 1.5×10^{-15}, whereas fractional curvature-redshift predicted by equation (11) for this density is 1.25×10^{-12}. Clearly, this is much larger. At γ ray frequencies, the electrons in the helium gas are effectively free and we can use equation (24) to compute the refractive index interaction length. For helium at STP, it is $L=0.077$ m, which is much less than curvature-redshift interaction length for these conditions of $X=11$ m. Hence, we do not expect to see any significant curvature-redshift in their results. Pound and Snyder did observe one-way frequency shifts but they were much smaller than curvature-redshift and could be explained by other aspects of the experiment.

However, the Pound and Snyder experiment provides a guide to a possible test for the existence of curvature-redshift. Because curvature-redshift has a different density variation to that for the inhibiting refractive index it is possible to find a density for which curvature-redshift is not inhibited. Although there is a slight advantage in using heavier gases than helium due to their higher atomic number to atomic weight ratio, their increased absorption to γ rays rules them out. Hence, we stay with helium and from equation (24) we can compute curvature-redshift interaction length to be

$$X = 10.8 \left(\frac{p_0}{p} \right)^{1/4} \text{ m}, \qquad (26)$$

where p is the pressure and p_0 is the pressure at STP. For the same gas the refractive index interaction length is

$$L = 0.077 \left(\frac{p_0}{p} \right) \text{ m}. \qquad (27)$$

It follows that the curvature-redshift will not be inhibited if $X<L$ or in this case, the pressure is less that $0.0014 p_0$ which is about 1 mm of Hg. For this pressure, we find that $X = 57$ m which requires that the apparatus must be much longer than 57 m. For argument let us take the length to be 100 m then the fractional redshift expected is 2.1×10^{-13} which is detectable. The experimental method would use a horizontal (to eliminate gravitational redshifts) tube filled with helium and with accurately controlled temperature. Then we would measure the redshift as a function of pressure. The above theory predicts that if it is free of inhibition then the redshift should be proportional to the square root of the pressure.

Alternatively, it may be possible to detect the secondary photons. For helium with a pressure of 1 mm Hg the expected frequency of the secondary radiation is about 100 kHz. The expected power from a 1 Cu source is about 5×10^{-22} W. Unfortunately, the secondary radiation could be spread over a fairly wide frequency band (equation (19)) which makes its detection somewhat difficult but it may be possible to detect the radiation with modulation techniques. Alternatively if γ rays of much shorter wavelength and power were used it may be possible to detect the secondary radiation in an experiment that did not try to measure the redshift.

3.6 Interactions for other particles

Since the focussing due to spacetime curvature applies to the quantum wave, it is expected that electrons and other particles would interact with curved spacetime in a manner similar to photons. The argument is the same up to equation (8) but now we have to allow for nonzero mass. The problem (not solved here) is to find a covariant expression that properly describes the energy-momentum loss to secondary particles and yet preserves the correct normalisation of the energy-momentum 4-vector. An alternate approach is to consider the motion in a local Minkowskian reference frame. In this case the loss equations (with P^0 denoting the energy component) are

$$\frac{dP^0}{dx} = \frac{\beta^2 P^0}{a_e},$$
$$\frac{dP^j}{dx} = \frac{P^j}{a_e}, \quad j = 1, 2, 3, \tag{28}$$

where β is the usual velocity ratio, a_e is the local radius of curvature for electrons and as required by normalisation, the conservation of proper mass, we have from equation (28)

$$\frac{dP^\alpha}{dx} P^\alpha = 0. \tag{29}$$

Noting that for a nonzero rest mass particle $V^\alpha V_\alpha = -1$. The radius of curvature a_e can be evaluated for the simple case of a uniform gas (or plasma) using equations (7) and (8) to get

$$a_e = \left\{ \frac{8\pi G}{c^2} \left[\left(\gamma^2 - \frac{1}{2} \right) \rho + \frac{p}{c^2} \left(\gamma^2 + \frac{1}{2} \right) \right] \right\}^{-\frac{1}{2}}, \tag{30}$$

where $\gamma = 1 / \sqrt{1 - \beta^2}$. Then with the further simplification of negligible pressure and with the material at rest the energy loss rate is

$$\boxed{\frac{1}{T} \frac{dT}{dx} = -\frac{1}{a_e} = -\left\{ \frac{8\pi G \rho \left(\gamma^2 - 1/2 \right)}{c^2} \right\}^{\frac{1}{2}} \beta^2,} \tag{31}$$

where $T = (\gamma - 1) mc^2$ is the kinetic energy. It shows that for nonzero rest mass particles, the energy loss rate has a strong dependence on velocity, and for extreme relativistic velocities, the fractional energy-loss rate is proportional to γ. Because of the strong velocity dependence, the energy loss rate for electrons will be much higher than that for nuclei in any plasma near thermal equilibrium. In addition, equation (31) shows that the energy loss rate has the same dependence on density as the energy loss rate for photons.

Since an electron interacts without being absorbed and re-emitted, we do not expect the same type of inhibition that applies to photons. Instead the electron slowly gets excited with the addition of energy (but no momentum), which it releases as low-energy photons when it interacts with some other particle. The need to preserve spin prevents it from emitting photons without the presence of another particle. In the cosmic medium, the most likely interactions are electro-magnetic scattering off other charged particles and the inverse-Compton effect off 3K background radiation photons. In high temperature plasma the electromagnetic (Rutherford) scattering is probably dominant since there will be many small angle deflections with large impact parameters. Thus the model for curvature-redshift of non-photon particles is one in which an excited electron emits most of its excitation energy as a low-energy photon during the scattering off another photon, electron or nucleus. Thus, we expect a secondary photon spectrum similar to equation (19).

3.7 Summary

Curvature-redshift is a process where a photon or any other particle interacts with curved spacetime. The interaction arises out of the focussing of waves by the geometry of spacetime with a positive curvature. The result of this interaction is that the particles lose energy at a rate proportional to the square root of the density of the medium through which they are passing. For a photon, the rate of energy loss is directly proportional to the photon energy and occurs without any deflection of the photon trajectory.

4 Derivation of curvature-pressure

4.1 Basic hypothesis

The hypothesis of curvature-pressure is that for particles that do not travel along geodesics there is a pressure generated by the non-geodesic motion. In this case, curvature-pressure acts on the matter (plasma) that is producing curved spacetime in such a way as to try to decrease the curvature. In other words, the plasma produces curved spacetime through its density entering the stress-energy tensor in Einstein's field equations. The magnitude of the curvature is an increasing function of the plasma density. Then the failure of the particles in the plasma to follow geodesics produces a pressure that acts back on the plasma to decrease the magnitude of the curvature and hence the density of the plasma.

4.2 Gravitation is not a force

The phrase 'gravitational force' is not only a popular expression but is endemic throughout physics. In particular, gravitation is classified as one of the four fundamental forces with its heritage going back to Newton's law of gravitation. I argue that the formulation of gravitation as a force is a misconception. In both Newtonian theory and General Relativity, gravitation is acceleration. The acceptance of the concept of gravitation as a force may explain why attempts to quantise the gravitational field have so far not been successful.

To begin let us examine the original Newtonian gravitation equation

$$m_I \mathbf{a} = \mathbf{F} = -\frac{GMm_G}{r^3}\mathbf{r}, \qquad (32)$$

where (following Longair, 1991) we identify m_I as the inertial mass of the test object, M as the active gravitational mass of the second object and m_G as the passive gravitational mass of the test object. The vector $\mathbf{a}$ is its acceleration and $\mathbf{r}$ is its displacement from the second object. This equation is usually derived in two steps: first, the derivation of a gravitational field and second, the force produced by that field on the test mass. By analogy with Coulombs law, the passive gravitational mass has a similar role to the electric charge.

However many experiments by Eötvös, Pekar & Fekete (1922), Dicke (1964), and Braginskii (1971) have shown that the passive gravitational mass is equal to the inertial mass to about one part in 10^{12}. The usual interpretation of the agreement is that they are fundamentally the same thing. However, an alternative viewpoint is that the basic equation is wrong and that the passive gravitational mass and the inertial mass should not appear in the equation. In this case the correct equation is

$$\mathbf{a} = -\frac{GM}{r^3}\mathbf{r}. \qquad (33)$$

Thus, the effect of gravitation is to produce accelerations directly; there is no force involved. Some might argue that since the two masses cancel the distinction is unimportant. On the other hand, I would argue that the application of Ockham's razor dictates the use of equation (33) instead of equation (32).

The theory of General Relativity is based on the principle of equivalence as stated by Einstein: 'All local, freely falling, non-rotating laboratories are fully equivalent for the performance of physical experiments.'

The relevance here is that it is impossible to distinguish between acceleration and a uniform gravitational field. This is illustrated by the geodesic motion of test particles. A test particle is one whose mass is small enough to have negligible effect on spacetime curvature. The important point is that the equations describing the geodesic of a test particle in General Relativity do not involve its mass, whereas in Newtonian gravity it is inserted and then cancelled by invoking the principle that the inertial mass is equal to the passive gravitational mass. When gravitation is considered as acceleration and not a force the passive gravitational mass is a spurious quantity that is not required by either theory.

4.3 A Newtonian model

A simple cosmological model using Newtonian physics in four-dimensional space illustrates some of the basic physics subsequently used to derive the features of curvature-pressure. The model assumes that the universe is composed of gas confined to the three-dimensional surface of a four-dimensional hypersphere. Since the visualisation of four dimensions is difficult let us suppress one of the normal dimensions and consider the gas to occupy the two-dimensional surface of a normal sphere. From Gauss's law (i.e. the gravitational effect of a spherical distribution of particles is identical to that of a point mass equal in value to the total mass situated at the centre of symmetry) the gravitational acceleration at the radius r of the surface is normal to the surface, directed inward and it has the magnitude

$$\ddot{r} = -\frac{GM}{r^2}, \tag{34}$$

where M is the total mass of the particles and the dots denote a time derivative. For equilibrium, and assuming all the particles have the same mass and velocity we can equate the radial acceleration to the gravitational acceleration and get the simple equation from celestial mechanics of

$$\frac{v^2}{r} = \frac{GM}{r^2}. \tag{35}$$

If there is conservation of energy, this stable situation is directly analogous to the motion of a planet about the sun. When there is a mixture of particles with different masses, there is an apparent problem. In general, particles will have a distribution of velocities and the heavier ones can be expected to have, on average, lower velocities. Thus, equilibrium radii will vary with the velocity of the particles. However, the basis of this model is that all particles are constrained to have the same radius regardless of their mass or velocity with the value of the radius set by the average radial acceleration. Thus for identical particles with a distribution of velocities we average over the squared velocities to get

$$\langle v^2 \rangle = \frac{GM}{r}. \tag{36}$$

If there is more than one type of particle with different masses then we invoke the precepts of section 4.2 and average over the accelerations to get the same result as equation (36). This is contrary to normal practice where the averaging would be over forces. The averaging over accelerations is a direct consequence of the arguments given in section 4.2 that gravitation is acceleration and not a force.

The effect of this balancing of the accelerations against the gravitational potential is seen within the shell as a curvature-pressure. If the radius r decreases then there is an increase in this curvature-pressure that attempts to increase the surface area by increasing the radius. For a small change in radius in a quasi-equilibrium process where the particle velocities do not change the work done by this curvature-pressure (two-dimensions) is $p_c dA$ and this must equal the gravitational force times the change in distance to give

$$p_c dA = \frac{GM^2}{r^2} dr, \qquad (37)$$

where $M = \sum m_i$ with the sum going over all the particles. Therefore, using equation (36) we can rewrite the previous equation in terms of the velocities as

$$p_c dA = \frac{M\langle v^2 \rangle}{r} dr. \qquad (38)$$

Now $dA/dr = 2A/r$, hence the two-dimensional curvature-pressure is

$$p_c = \frac{M\langle v^2 \rangle}{2A}. \qquad (39)$$

Thus in this two-dimensional model the curvature-pressure is like the average kinetic energy per unit area. This simple Newtonian model provides a guide as to what the curvature-pressure would be in the full general relativistic model.

4.4 General Relativistic model

In deriving a more general model in analogy to the Newtonian one, we first change $dA/dr=2A/r$ to $dV/dr=3V/r$ and secondly we include the correction γ^2 needed for relativistic velocities. The result is

$$p_c = \frac{M\langle \gamma^2 \beta^2 \rangle c^2}{3V} = \frac{\langle \gamma^2 -1 \rangle Mc^2}{3V}, \qquad (40)$$

where $\gamma = 1/\sqrt{1-\beta^2}$ and $\beta = v/c$. Now we expect to be dealing with fully ionised high temperature plasma with a mixture of electrons, protons, and heavier ions where the averaging is done over the accelerations. Define the average density by $\rho = M/V$ then the cosmological curvature-pressure is

$$\boxed{p_c = \frac{1}{3}\langle \gamma^2 - 1 \rangle \rho c^2}. \tag{41}$$

In effect, my hypothesis is that the cosmological model must include this curvature-pressure as well as thermodynamic pressure. Note that although this has a similar form to thermodynamic pressure it is quite different. In particular, it is proportional to an average over the squared velocities and the thermodynamic pressure is proportional to an average over the kinetic energies. This means that, for plasma with free electrons and approximate thermodynamic equilibrium, the electrons will dominate the average due to their much larger velocities. From a Newtonian point of view, curvature-pressure is opposed to gravitational mutual acceleration. In other words, the plasma produces curved spacetime through its density entering the stress-energy tensor in Einstein's field equations. Then the failure of the particles in the plasma to follow geodesics produces a pressure whose reaction is the curvature-pressure acting to decrease the magnitude of the curvature and hence decrease the density of the plasma.

For high temperature plasma in equilibrium, the Jüttner distribution can be used to evaluate the curvature-pressure. For a gas with temperature T and for molecules with mass m, de Groot, Leeuwen & van Weert (1980) showed that

$$\gamma^2(\alpha) = 3\alpha K_3(1/\alpha)/K_2(1/\alpha), \tag{42}$$

where $\alpha = kT/mc^2$ and $K_n(1/\alpha)$ are the modified Bessel functions of the second kind (Abramowitz & Stegun 1968). For small, α this has the approximation

$$\gamma^2(\alpha) = 1 + 3\alpha + 152\alpha^2 + 458\alpha^3 + \ldots. \tag{43}$$

For a Maxwellian (non-relativistic) distribution, the first two terms are exact and the α^2 term is the first term in the correction for the Jüttner distribution.

4.5 Local curvature-pressure

For the universe, the calculation of curvature-pressure is simple because of the constant curvature and homogeneous medium. However, for a localised region such as a star with inhomogeneous medium and curvature the calculation is much more difficult. We start with the premise that it is the non-geodesic motion of particles that reacts back on the material producing the curvature by producing a pressure that tends to reduce the curvature. The problem is that the calculation of the curvature at any point requires the integration of Einstein's equations of General Relativity. Then if the particles' lack of geodesic motion produces a reaction force, the problem is to determine how that reaction force is apportioned amongst the matter that produces the curvature.

One approach that is valid for most astrophysical applications where the spacetime curvature is small is to use the Newtonian approximation. Let a, be the effective radius of curvature of the four dimensional space where the particles' motion is dominated by non-gravitational forces. Then the premise is that this motion produces acceleration due to curvature (assuming for the moment that there is only one type of particle)

$$g_c = \frac{\langle v^2 \rangle}{a} \qquad (44)$$

where the bar denotes an averaging over all the velocities. Now consider a spherically symmetric distribution of gas. If the distribution is static, the central gravitational attraction is balanced by some pressure p_g, so that

$$\frac{dp_g}{dr} = -\rho(r)g, \qquad (45)$$

where $\rho(r)$ is the density at radius r and g is the gravitational acceleration at r. Similarly, we define a curvature-pressure by

$$\frac{dp_c}{dr} = -\rho(r)g_c. \qquad (46)$$

However, if there is a mixture of particles there is an important difference. Because electrons have a much lighter mass than ions the velocity average for mixed particles (provided the gas is ionized) will be dominated by the electrons and the appropriate density to use in equation (46) is that for the electrons. Now the curvature radius a, is given by equation (10), and for a gas with relativistic particles we put

$$\langle v^2 \rangle = \langle \gamma^2 - 1 \rangle c^2. \tag{47}$$

We need to include a factor of one third because only the velocity component orthogonal to the direction of the acceleration is relevant. Then the curvature reaction acceleration is

$$g_c = \frac{1}{3}\langle (\gamma^2 - 1)\sqrt{\rho(r)} \rangle c^2 \sqrt{\frac{8\pi G}{c^2}}, \tag{48}$$

and

$$\frac{dp_c}{dr} = -\frac{1}{3}\langle (\gamma^2 - 1)\sqrt{\rho(r)} \rangle c^2 \sqrt{\frac{8\pi G}{c^2}} \rho(r). \tag{49}$$

Since the hypothesis is that this curvature-pressure is a reaction to the accelerations produced by the gas at radius r, the averaging over velocities must be over all the gas that is being accelerated. By Gauss's law and symmetry this is the gas with radii greater than r thus we get

$$\langle (\gamma^2 - 1)\sqrt{\rho(r)} \rangle = \int_r^\infty N(\hat{r})\hat{r}^2 (\gamma^2 - 1)\sqrt{\rho(\hat{r})} d\hat{r} \Big/ \int_r^\infty N(\hat{r})\hat{r}^2 d\hat{r}, \tag{50}$$

where $N(r)$ is the particle number density. Now for plasmas where the temperatures less than about 10^8K we can use equation (43) to get

$$\frac{1}{3}\langle \gamma^2 - 1 \rangle = \frac{kT}{m_e c^2}. \tag{51}$$

Hence the working equation for local curvature-pressure is

$$\boxed{\frac{dp_c}{dr} = -k\langle T(r)\sqrt{\rho(r)} \rangle \sqrt{\frac{8\pi G}{c^2}} \rho(r)}, \tag{52}$$

where

$$\langle T(r)\sqrt{\rho(r)} \rangle = \int_r^\infty N_e(\hat{r})\hat{r}^2 T(\hat{r})\sqrt{\rho(\hat{r})} d\hat{r} \Big/ \int_r^\infty N_e(\hat{r})\hat{r}^2 d\hat{r}, \tag{53}$$

and $N_e(r)$ is the electron number density.

Although this derivation has assumed plasma where the dominant particle forces are electromagnetic, the hypothesis is applicable to other material where the strong nuclear force dominates. In highly condensed matter, the strong nuclear forces prevent the particles from following geodesics; therefore, we expect curvature-pressure to be present. A theory of curvature-pressure in a very dense medium where quantum mechanics dominates and where General Relativity may be required is needed to develop this model. Nevertheless, without such a theory, we expect the pressure to be proportional to the local gravitational acceleration and an increasing function of the temperature of the particles. Thus, we might expect a curvature-pressure that would resist a hot compact object from collapsing to a black hole. Because of the energy released during collapse, it is unlikely for a cold object to stay cold enough to overcome the curvature-pressure and collapse to a black hole. This is the explanation for the earlier comment (section 1.1) that black holes are unlikely to occur.

5 The curvature-cosmological model

5.1 Introduction

Curvature-cosmology can now be derived by incorporating curvature-redshift and curvature-pressure into the equations of General Relativity. This is done by using homogeneous isotropic plasma as a model for the real universe. The General Theory of Relativity enters through the Friedmann equations for a homogeneous isotropic gas. Although such a model is simple compared to the real universe, the important characteristics of curvature-cosmology can be derived by using this model. The first step is to obtain the basic relationship between the density of the gas and the radius of the universe. The inclusion of curvature-pressure is not only important in determining the basic equations but it also provides the necessary means of making the solution static and stable. Then it is shown that the effect of curvature-redshift is to produce a redshift that is a function of distance, and the slope of this relationship is (in the linear limit of small distances) the Hubble constant.

The next step is to investigate the geometry of curvature-cosmology and to derive the equations for luminosity, angular-size, and surface-brightness as a function of redshift. The key equations that define the curvature-cosmology geometry are equation (69), which defines the radius of the universe in terms of the Hubble constant, and equation (71) which relates the distance variable to the redshift.

Finally, the cosmologically important auxiliary topics of entropy, Olber's paradox, nuclear abundances, and black holes are discussed in the context of curvature-cosmology.

5.2 The homogenous isotropic model

Following normal practice, the derivation proceeds by considering a cosmological model for uniform plasma. The first-order model considers the universe to be a gas with uniform density and complications such as density fluctuations, galaxies, and stars are ignored. In some cases, we may consider a second order model that incorporates density variations. In addition, we assume (to be verified later) that the gas is at high temperature and is fully ionised plasma. Because of the high symmetry, the appropriate metric is the one that satisfies the equations of General Relativity for a homogeneous, isotropic gas. This metric was first discovered by A. Friedmann and fully investigated by H. P. Robertson and A. G. Walker. The Robertson-Walker metric for a space with positive curvature can be written (Rindler 1977, p320) as

$$ds^2 = c^2 dt^2 - [R(t)]^2 \left[\frac{dr^2}{1-r^2} + r^2 \left(d\theta^2 + \sin^2(\theta) d\varphi^2 \right) \right], \quad (54)$$

where ds is the interval between events, dt is time, $R(t)dr$ is the co-moving increment in radial distance, $R(t)$ is the radius of curvature and R_0 is the value of $R(t)$ at the present epoch.

5.2.1 The Friedmann equations

Based on the Robertson-Walker metric, the Friedmann equations for the homogeneous isotropic model with constant density and pressure are (Longair 1991)

$$\ddot{R} = -\frac{4\pi G}{3}\left(\rho + \frac{3p}{c^2}\right)R + \frac{1}{3}\Lambda R,$$
$$\dot{R}^2 = \frac{8\pi G}{3}\rho R^2 - c^2 + \frac{1}{3}\Lambda R^2 \quad (55)$$

where R is the radius, ρ is the proper density, p is the thermodynamic pressure, G is the Newtonian gravitational constant, Λ is the cosmological constant, c is the velocity of light and the superscript dots denote time derivatives. Working to order of m_e/m_p thermodynamic pressure may be neglected but not curvature-pressure. How to include curvature-pressure is not immediately obvious. The thermodynamic pressure appears only as a relativistic correction to the inertial mass density whereas curvature-pressure is closer in spirit to the cosmological constant. My solution is to include curvature-pressure (with a negative sign) with the thermodynamic pressure and to set the cosmological constant to zero. Including curvature-pressure from equation (41) equations (55) are now

The homogenous isotropic model

$$\ddot{R} = -4\pi G\rho\left[1-\langle\gamma^2-1\rangle\right]R,$$

$$\dot{R}^2 = \frac{8\pi G\rho}{3}R^2 - c^2 \qquad (56)$$

Clearly there is a static solution if $\langle\gamma^2-1\rangle=1$, in which case $\ddot{R}=0$. The second equation, with $\dot{R}=0$ provides the radius of the universe which is given by

$$R = \sqrt{\frac{3c^2}{8\pi G\rho}} = \sqrt{\frac{3c^2}{8\pi GM_H N}}. \qquad (57)$$

Thus, the model is a static cosmology with positive curvature. Although the geometry is similar to the original Einstein static model, this curvature-cosmology differs in that it is stable. The basic instability of the static Einstein model is well known (Tolman 1934, Ellis 1984). On the other hand, the stability of curvature-cosmology is shown by considering a perturbation ΔR, about the equilibrium position. Then the perturbation equation is

$$\Delta\ddot{R} = \frac{c^2}{8\pi R_0}\left(\overline{\frac{d\gamma^2}{dR}}\right)\Delta R, \qquad (58)$$

and since for any realistic equation of state for the cosmic plasma, the average velocity will decrease as R increases. Then the right hand side is negative, showing that the result of a small perturbation is for the universe return to its equilibrium position. Thus, the curvature-cosmology model is intrinsically stable. Of theoretical interest is that equation (58) predicts that oscillations could occur about the equilibrium position.

5.2.2 Temperature of the universal plasma

One of the most remarkable results of curvature-cosmology is that it predicts the temperature of the cosmic plasma from fundamental constants. That is the predicted temperature is independent of the density and independent of any other characteristic of the universe. For a stable solution to equation (56) we need that $\langle\gamma^2-1\rangle=1$, (i.e. $\langle\gamma^2\rangle=2$) where the average is taken over the electron and nucleon number densities, that is for equal numbers of electrons and protons

$$\langle \gamma^2 \rangle \cong 0.5 \langle \gamma_e^2 + \gamma_p^2 \rangle, \tag{59}$$

where the terms on the right are for electrons and protons. Provided the temperatures are small enough for the proton's kinetic energy to be much less than its rest mass energy, we can put $\langle \gamma_p^2 \rangle = 1$ and thus for pure hydrogen, the result is $\langle \gamma_e^2 \rangle = 3$. Using a more realistic composition that has 8.5% by number (Allen 1976) of helium we find that $\langle \gamma_e^2 \rangle = 2.927$. Hence using equation (42) the predicted electron temperature is 2.56×10^9 K. For this temperature $\langle \gamma_p^2 \rangle = 1.0007$. This shows that the temperature is low enough to justify the assumption made earlier, that the proton's kinetic energy is much smaller than its rest mass energy.

To recapitulate the stability of curvature-cosmology requires that $\ddot{R} = 0$. This requires that the plasma has the precise temperature that makes $<\gamma^2-1>=1$. The basis for this result is that curvature-pressure exists and critical to its derivation is the averaging over accelerations and not over forces. This is where the assertion that gravitation is acceleration and is not a force is important.

5.2.3 Hubble constant: theory

If the gravitational curvature-redshift is to explain the Hubble redshift then, to be consistent, the stress-energy-momentum tensor for curvature-cosmology should be used in deriving curvature-redshift. Thus curvature-pressure must be included in equation (55), and since $<\gamma^2>=2$ its value from equation (41) is $p_c = -\rho c^2/3$. Hence the relative energy loss rate is

$$\frac{1}{E}\frac{dE}{ds} = -\frac{1}{a} = -\left(\frac{16\pi G M_H N}{3c^2}\right)^{1/2}. \tag{60}$$

$$= -4.56 \times 10^{-27} N^{1/2} \mathrm{m}^{-1}$$

Integration of equation (60) (for a homogeneous medium) gives

$$E_0 = E_e e^{-x/a}, \tag{61}$$

where the emitted energy is E_e, the final energy is E_0, and x is the distance the photon has travelled. The usual redshift parameter z is defined in terms of the wavelengths as

$$z = \frac{\lambda_0}{\lambda_e} - 1 = \frac{v_e}{v_0} - 1 = \frac{E_e}{E_0} - 1 = \exp\left(\frac{x}{a}\right) - 1 \quad (62)$$

$$z = \frac{x}{a} + \frac{1}{2}\left(\frac{x}{a}\right)^2 + \frac{1}{6}\left(\frac{x}{a}\right)^3 + \ldots$$

If we define the Hubble constant to be the coefficient of the linear term then $H = c/a$ where the velocity of light is included to give H its usual scaling of inverse time.

$$\boxed{\begin{aligned} H &= \left(\frac{16\pi G M_H N}{3}\right)^{1/2} = 1.368 \times 10^{-18} N^{1/2}.s^{-1} \\ &= 42.21 N^{1/2} \text{ km.s}^{-1}.\text{Mpc}^{-1} \end{aligned}} \quad (63)$$

For a non-cosmological application, that is for distances that do not include the cosmic plasma, and where temperatures are much lower than the cosmic temperature, the appropriate expression for the redshift is equation (11) and not equation (60).

5.3 Geometry of curvature-cosmology

The Robertson-Walker metric shown in equation (55) is not in the simplest form that explicitly shows the geometry. Following D'Inverno (1992) we can introduce a new variable χ, where

$$r = R \sin \chi, \quad (64)$$

and the new metric is

$$ds^2 = c^2 dt^2 - R^2 \left[d\chi^2 + \sin^2 \chi \left(d\theta^2 + \sin^2 \theta d\phi^2 \right) \right]. \quad (65)$$

In this metric the distance travelled by a photon is $R\chi$, and since the velocity of light is a universal constant the time taken is $R\chi/c$. There is a close analogy to motion on the surface of the earth with radius R. Light travels along great circles and χ is the angle subtended along the great circle between two points.

The geometry of this curvature-cosmology is that of a three-dimensional surface of a four-dimensional hypersphere. It is identical to that for Einstein's static universe and is the same as that for the Big-Bang model without expansion. For a static universe, there is no ambiguity in the definition of distances and times. One can use a universal cosmic time and define distances in light travel times or any other convenient measure. For this geometry the area of a sphere with radius $R\chi$ is given by

$$A(r) = 4\pi R^2 \sin^2(\chi). \tag{66}$$

The surface is finite and χ can vary from 0 to π. Integration of this equation with respect to χ gives the volume V, namely,

$$V(r) = 2\pi R^3 \left[\chi - \frac{1}{2}\sin(2\chi) \right]. \tag{67}$$

Clearly the maximum volume is $2\pi^2 R^3$ and using equation (60)

$$R = \sqrt{\frac{3c^2}{8\pi G M_H N}} = 3.100 \times 10^{26} N^{-\frac{1}{2}} \text{ m} = 10.05 N^{-\frac{1}{2}} \text{ Gpc}. \tag{68}$$

Then for example with $N=2$, $R = 4.38 \times 10^{26}$ m and the volume is 1.66×10^{81} m^3. The mass for this case is 5.56×10^{54} kg.

The next step is to combine curvature-redshift prediction equation (60) with that for the cosmological model equation (57) to get

$$\boxed{R = \sqrt{2}a = \frac{\sqrt{2}c}{H}}. \tag{69}$$

Therefore, we can include the redshift into the geometry to get

$$z = \exp\left\{\frac{Hx}{c}\right\} - 1 = \exp\left\{\frac{HR\chi}{c}\right\} - 1 = \exp(\sqrt{2}\chi) - 1, \tag{70}$$

and after inversion

$$\boxed{\chi = \ln(1+z)/\sqrt{2}}. \tag{71}$$

This is the fundamental relationship between z and χ. We also have the useful relation

$$\frac{d\chi}{dz} = \frac{1}{\sqrt{2}(1+z)}. \tag{72}$$

If there is a class of objects having a uniform density n_0 then the expected distribution as a function of z is

$$\frac{dn}{dz} = n_0 \frac{dV(z)}{dz} = \pi n_0 \left(\frac{2c}{H}\right)^3 \frac{\sin^2(\chi)}{1+z}. \tag{73}$$

The predicted number distribution (without selection) for any uniformly distributed objects is shown in Figure 1 for both

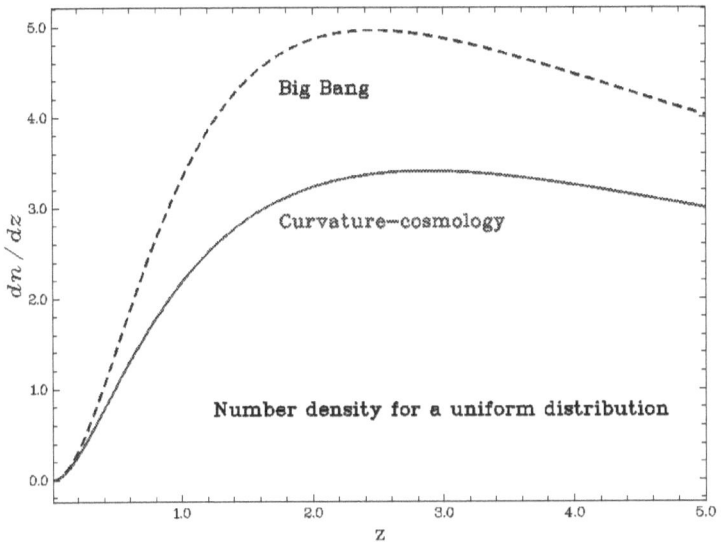

Figure 1: Number density as a function of redshift.

curvature-cosmology and Big-Bang cosmology with $\Omega_M=0.3$ (see below for more details) where the density is referred to the co-moving volume. This is the predicted number-density distribution as a function of redshift for any type of object for scales that are larger than any local clustering. Naturally, the observed distribution of objects such as quasars will be significantly modified by selection effects.

5.3.1 Geometry : conclusion

Since the geometry of curvature-cosmology does not involve a time coordinate, it is much simpler than that for Big-Bang cosmology. The key equations define the curvature-cosmology geometry are equation (69) which defines the radius of the universe in terms of the Hubble constant and equation (71) which defines the distance variable χ in terms of the redshift parameter z. We now examine some topics that are relevant to all cosmologies.

5.4 Luminosities and magnitudes

Let a source have a luminosity $L(\nu)$ (W.Hz^{-1}) at the emission frequency ν. Then if energy is conserved, the observed flux density ($W.m^{-2}Hz^{-1}$) at a distance parameter r is the luminosity divided by the area, which is

$$S(v)dv = \frac{L(v)dv}{4\pi \{R\sin(\chi)\}^2}. \qquad (74)$$

Note that the light travel distance is $R\chi$ and not the distance parameter r. However, because of curvature-redshift there is an energy loss such that the received frequency v_0 is related to the emitted frequency v_e by equation (62). Including this effect the result is

$$S(v_0)dv_0 = \frac{L(v_e)dv_e}{4\pi (R\sin(\chi))^2 (1+z)}. \qquad (75)$$

The apparent magnitude is defined as $m=-2.5\log(S)$ where the base of the logarithm is 10 and the constant 2.5 is exact. Since the absolute magnitude is the apparent magnitude when the object is at a distance of 10 pc (3.0857×10^{17} m), the flux density at 10 pc is

$$S_{10}(v_0)dv_0 = \frac{L(v_0)dv_0}{2\pi(10pc)^2}, \qquad (76)$$

where because 10 pc is negligible compared to R, approximations have been made. The flux density ratio is

$$\frac{S(v_0)}{S_{10}(v_0)} = \left\{\frac{10pc}{R\sin(\chi)}\right\}^2 \left\{\frac{L(v_e)dv_e}{L(v_0)dv_0}\right\}\left\{\frac{1}{1+z}\right\}. \qquad (77)$$

Defining M as the absolute magnitude and putting $v_e=(1+z)v_0$ we get for the distance modulus $(m-M)$

$$m - M = -2.5\log\left(\frac{S(v_0)}{S_{10}(v_0)}\right)$$

$$= 5\log\left\{\frac{R\sin(\chi)}{10pc}\right\} - 2.5\log\left\{\frac{L((1+z)v_0)(1+z)dv_0}{(1+z)L(v_0)dv_0}\right\} \qquad (78)$$

$$= 5\log\left\{\frac{R\sin(\chi)}{10pc}\right\} + K_z(v_0) + 2.5\log(1+z)$$

where the K-correction is (Rowan-Robertson 1985, Peebles 1993, and Hogg et al. 2002, who provides a thorough discussion)

$$K_z(v_0) = -2.5\log\left\{\frac{(1+z)L((1+z)v_0)}{L(v_0)}\right\}, \qquad (79)$$

For broadband filters, the luminosities are replaced by appropriate integrals over the product of the filter transfer functions and the luminosity spectrum. Note that it is traditional to include the bandwidth ratio (1+z) in the K-correction. This is why equation (79) includes a term 2.5log(1+z) in order to cancel the one included in the K-correction. Furthermore, we can use equation(69) to replace R by H since

$$\frac{R}{\sqrt{2}} = \frac{c}{H} = \frac{2.998}{h} \text{ Gpc}, \qquad (80)$$

Where h is the reduced Hubble constant ($H=100h$ km.s^{-1}.Mpc^{-1}). Hence, we get

$$\boxed{\begin{array}{l} m - M = 5\log\left[\sqrt{2}\sin(\chi)\right] + 2.5\log(1+z) + K_z(v_0) \\ -5\log(h) + 42.384 \end{array}}, \qquad (81)$$

where from equation (71) $\chi = \ln(1+z)/\sqrt{2}$. The right hand side of equation (81) without the K-correction is called the distance modulus (*DM*).

For comparison Big-Bang cosmology has a much more complex expression that uses the dimensionless energy density parameters: Ω_M the mass energy density; Ω_Λ the cosmological energy density; and Ω_k the curvature energy density which are defined by

$$\Omega_M = \frac{4\pi G \rho_0}{3H_0^2}, \quad \Omega_\Lambda = \frac{\Lambda c^2}{3H_0^2}, \qquad (82)$$

where Λ is the cosmological constant and ρ_0 is the current density. Then the distance modulus *DM* can be written as

$$DM = 5\log(D_L) + 5\log(H_0) - 25, \qquad (83)$$

where D_L is the luminosity-distance, H_0 has the units km.s^{-1}.Mpc^{-1} and the constant, 25, comes from the definition that $M=m$ at 10 pc. Then following Goobar & Perlmutter (1995) (with corrections from Perlmutter et al. 1997) and Hogg (1999), the function $S(x)$ is shown in Table 2 for ranges of the mass-density parameter Ω_M and the cosmological-constant energy-density parameter Ω_Λ. The luminosity-distance is give by

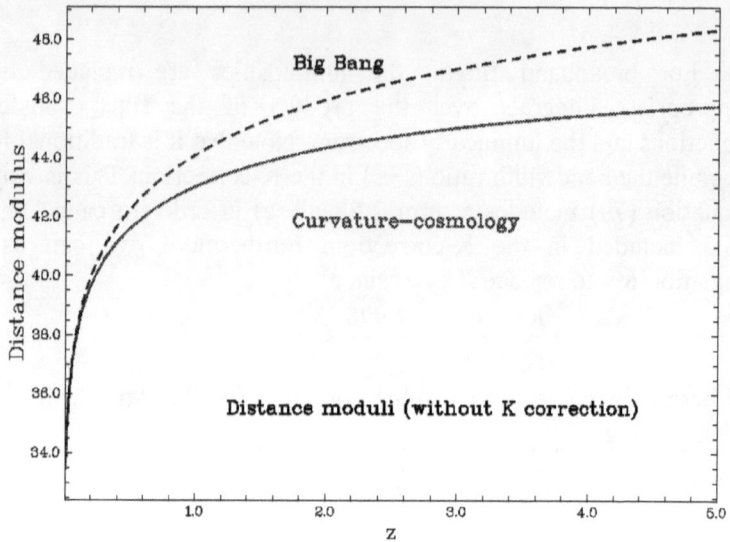

Figure 2: Distance moduli (*DM*) for Big-Bang and Curvature cosmologies.

$$D_L = \frac{c(1+z)}{H_0\sqrt{|\kappa|}} S\left(\sqrt{|\kappa|} \int_0^z \left[(1+z')^3 \Omega_M + (1+z')^2 \Omega_k + \Omega_\Lambda\right]^{-\frac{1}{2}} dz'\right). \quad (84)$$

For the commonly used value $\Omega_M + \Omega_\Lambda = 1$ this becomes

$$D_L = \frac{c(1+z)}{H_0}\left(\int_0^z \left((1+z')^3 \Omega_M + \Omega_\Lambda\right)^{-\frac{1}{2}} dz'\right). \quad (85)$$

Table 2: Big-Bang luminosity-distance parameters.

Range	$S(x)$
$\Omega_M + \Omega_\Lambda > 1$	$S(x) = \sin(x)$
$\Omega_M + \Omega_\Lambda = 1$	$S(x) = 1$
$\Omega_M + \Omega_\Lambda < 1$	$S(x) = \sinh(x)$

The bolometric distance modulus is given by equation (81) for curvature-cosmology and for Big-Bang cosmology it is

$$5\log\left((1+z)\int_0^z \left((1+z')^3 \Omega_M + 1 - \Omega_M\right)^{-\frac{1}{2}} dz'\right) \ .$$

These distance moduli are shown in Figure 2. In curvature-cosmology the distance modulus is always less than it is for Big-

Bang cosmology with the parameters chosen above. For the same absolute magnitude, an object will appear to be brighter in curvature-cosmology than it does in Big-Bang cosmology.

5.5 Angular size

One of the simplest tests of cosmological models is to measure the angular size of an object as a function of redshift. If d is the transverse linear size then with a small angle approximation the observed angular size, θ, is (with $\chi = \ln(1+z)/\sqrt{2}$)

$$\theta = \frac{d}{R\sin(\chi)}. \tag{86}$$

If we could observe to high enough redshifts, we would find that the angular size decreases to a minimum at $z=8.2$ (where $\chi=\pi$) and then starts to increase. For θ measured in arc seconds and with d measured in parsecs equation (86) becomes

$$\log(d) = \log(\theta'') + \log(\sqrt{2}\sin(\chi)) + \log(h) + 5.209. \tag{87}$$

Since the distance (e.g. light time) to an object is $R\chi$, it is apparent that in the limit for small distances equation (86) reduces to the Euclidean equation $\theta=d/R\chi$. Since there are no known objects that have well-defined sizes and can be seen at cosmological distances, the main application of the test relies on examining the statistical properties of large radio sources.

5.6 Surface brightness

The surface brightness of an object is the power produced per unit surface area. In a Euclidean universe, the reduction in intensity due to the inverse square law is exactly balanced by the change in observed area so that the bolometric surface brightness of an object is constant. For curvature-cosmology (as for any static-tired-light cosmology), the only difference from the Euclidean case is that there is a factor of $(1+z)^{-1}$ in order to allow for energy loss. For an expanding universe, the surface brightness varies as $(1+z)^{-4}$ (Peebles 1993) where one factor of $(1+z)$ comes from the decrease in energy of each photon due to the redshift, another factor comes from the decrease in rate of their arrival and two factors come from the apparent decrease in area due to aberration. Thus, in principle observations of surface brightness can make a powerful test for the difference between expanding and static tired-light cosmology models. In practice, surface brightness is measured in finite frequency bands thus requiring the additional complication of the K-correction. As shown later a major problem in using surface brightness is the lack of suitable objects. Even relatively well-understood objects such as elliptical galaxies have difficulties in that their surface brightness distribution is not uniform and varies with their absolute size.

5.7 Auxiliary topics

There are many topics that are part of cosmology but do not directly involve distance-redshift relationships. They are considered here either because they do not lend themselves to direct observation or, like Olber's paradox, the observations are trivial.

The properties of entropy in a static universe differ fundamentally from those in an expanding universe. The often-asserted proposition that the entropy of the universe is increasing is clearly at odds with the notion of a static universe. Similarly the most basic cosmological question of 'why is the sky dark at night', Olber's paradox, requires examination.

Another fundamental topic is the abundance of the light elements. In Big-Bang cosmology these are called the 'primordial abundance' since the light elements were produced in the very early universe. In curvature-cosmology, the primordial abundance refers to the abundance in the cosmic plasma from which the galaxies are formed. The final topic to be discussed here is the important subject of black holes.

5.7.1 Entropy

Consider a small body in circular motion about a mass M. Then the Newtonian gravitational acceleration will equal the centripetal acceleration to get

$$\frac{v^2}{r} = \frac{GM}{r^2} \text{ or }$$
$$v^2 = \frac{GM}{r}$$
(88)

If we multiply by the mass of the body, we can get a simple relationship between its kinetic energy and its potential energy.

This is a specific example of a more general statistical relationship called the 'virial theorem' (Goldstein 1980). Consider a stellar cluster or an isolated cloud of gas in which collisions are negligible or elastic. In either case the virial theorem states that the average kinetic energy K, is related to the average potential energy V, by the equation
$$V = V_0 - 2K, \qquad (89)$$
where V_0 is the potential energy when there is zero kinetic energy. Let U be the total energy then
$$U = K + V = V_0 - K. \qquad (90)$$

Thus, we get the somewhat paradoxical situation that since V_0 is constant; an increase in total energy can cause a decrease in kinetic energy. For example, consider a stellar cluster or a gas cloud that is in equilibrium. Equation (90) shows that an increase in total energy will lead to a decrease in kinetic energy. This happens because the average potential energy has increased by approximately twice as much as the loss in kinetic energy. Since the temperature is proportional to (or at the least a monotonic increasing function of) the average kinetic energy it is apparent that an increase in total energy leads to a decrease in temperature. This explains the often-quoted remark that a self-gravitationally bound gas cloud has a negative specific heat capacity. Thus, when gravity is involved the whole construct of entropy needs to be reconsidered.

One of the common statements of the second law of thermodynamics is that (Longair 1984): *The energy of the universe is constant: the entropy of the Universe tends to a maximum*, or (Feynman, Leighton & Sands 1965): *the entropy of the universe is always increasing*.

Now the normal proof of the second law considers the operation of reversible and non-reversible heat engines working between two or more heat reservoirs. If we use a self-gravitating gas cloud as a heat reservoir then we will get quite different results since the extraction of energy from it will lead to an increase in its temperature. Thus if the universe is dominated by gravity the second law of thermodynamics needs reconsideration. In addition, it should be noted that we cannot have a shield that hides gravity.

To put it another way there is no adiabatic container that is beyond the influence of external gravitational fields.

Consider a thought experiment in which there are a large number of molecules, at rest, uniformly spread in space except that at the origin there is one extra molecule. Because of the symmetry and low velocities, Newtonian physics is adequate. We also assume that the molecules have only elastic collisions and that there are no other energy loss mechanisms present, such as radiation. Because of the extra molecule, there will be a (very) slow acceleration of the molecules towards the origin. Eventually some will interact, the initial symmetry of the radial motion will be destroyed, and the molecules will form a self-gravitating cloud with a continual influx of more distant molecules. Clearly, what is happening is that some of the potential energy due to the initial configuration has been turned into kinetic energy and the average temperature of the molecules has been increased. The interesting question is what has happened to the entropy. It seems that we have gone from a state of extreme order to one of disorder, which would imply an increase in entropy. However, if we suddenly reverse the direction of time the molecules will have the reverse interactions and will return to their initial positions. Thus, the process is reversible: there is no change in entropy. Now if we include quantum mechanics, or even allow a minute randomness in the molecular collisions, the process is no longer reversible but the difference in the end product could not be detected without knowing the history of the particles.

This discussion shows that in a static finite universe dominated by gravity simple discussions of the second law of thermodynamics can be misleading. The presence of gravity means that it is impossible to have an isolated system. To be convincing any proof of the second law of thermodynamics should include the universe and its gravitational interactions in the proof.

5.7.2 Olber's Paradox

For curvature-cosmology, Olber's Paradox is not a problem. Curvature-redshift is sufficient to move distant starlight out of the visible band. Visible light from distant galaxies is shifted into the infrared where it is no longer seen.

Of course, with a finite universe, there is the problem of conservation of energy and why we are not saturated with very low frequency radiation produced by curvature-redshift. These low-energy photons are eventually absorbed by the cosmic plasma. Everything is recycled. The plasma radiates energy into the microwave background radiation and into X-rays. The galaxies develop from the cosmic plasma and pass through their normal evolution. Eventually all their material is returned to the cosmic plasma. Note that very little, if any, is locked up into black holes. Curvature-pressure causes most of the material from highly compact objects to be returned to the surrounding region as high-velocity jets.

5.7.3 Nuclear abundances

One of the successes of Big-Bang cosmology is in its explanation of the primordial abundances of the light elements. Since the proposed curvature-cosmology is static, there must be another method of getting the 'primordial' abundances of light elements. In curvature-cosmology, the primordial abundance refers to the abundance in the cosmic plasma from which the galaxies are formed.

The first point to note is that in curvature-cosmology the predicted temperature of the cosmic plasma is 2.56×10^9 K at which temperature nuclear reactions can proceed. It is postulated that there is a continuous recycling of material from the cosmic plasma to galaxies and stars and then back to the plasma. Because of the high temperature, nuclear reactions will take place whereby the more complex nuclei are broken down to hydrogen, deuterium, and helium. Although this cycling can take many billions of years the very low density of the plasma means that the cycle time may not be long enough for the nuclei densities to achieve statistical equilibrium. In addition, the major reactions required are the breaking down of heavier nuclei to lighter ones and not those that construct nuclei. It is through the interactions of cosmic plasma in curvature-cosmology that 'primordial' abundances that are seen in early galaxy formation are produced.

5.7.4 Black holes and Jets

The existence of curvature-pressure provides a mechanism that could prevent the collapse of a compact object into a black hole. At sufficiently high densities, the electrons and protons combine into neutrons and the electromagnetic force is no longer important in preventing the particles from following geodesics. However, the strong nuclear force is present and will prevent geodesic motion, hence producing curvature-pressure. A theory of curvature-pressure in a very dense medium where quantum mechanics dominates is needed to develop this model. Nevertheless, without a full theory we can assume that curvature-pressure will depend on the local gravitational acceleration and it will be an increasing function of the temperature of the particles. Thus, we might expect a curvature-pressure that would resist a hot compact object from collapsing to a black hole. Because of the energy released during collapse it is unlikely for a cold object to stay cold enough to overcome the curvature-pressure and collapse to a black hole.

What is expected is that the final stage of gravitational collapse is a very dense object, larger than a black hole but smaller than a neutron star. This compact object would appear very much like a black hole and would have most of the characteristics of black holes. Such objects could have large masses and be surrounded by accretion discs. Thus, many of the observations that are thought to show the presence of black hole could equally show the presence of these compact objects. However, there is one observational difference in that many of the mass estimates of black holes come from observations of redshifts from nearby stars. Since part or most of these redshifts may be due to curvature-redshift in the surrounding gas, these mass estimates may need to be revised.

If the compact object is rotating there is the tantalising prospect that curvature-pressure may produce the emission of material in two jets along the spin axis. This could be the 'jet engine' that produces the astrophysical jets seen in stellar-like objects and in many huge radio sources. Currently there are no accepted models for the origin of these jets. One suggestion is that they due to the twisting of a magnetic fields threading a differentially rotating accretion disk which acts to magnetically extract angular momentum and energy from the disk. There are two possibilities: hydro-magnetic outflows, which have a significant mass flux and have energy and angular momentum carried by both the matter and the electromagnetic field and, Poynting outflows, where the mass flux is negligible and energy and angular momentum are carried predominantly by the electromagnetic field.

The postulate here is that the jets are a property of the compact object and do not come from the accretion disk. The spinning object provides the symmetry necessary to generate two jets and curvature-pressure provides the force that drives the jets. This mechanism is applicable to both stellar and galactic size structures.

5.7.5 Large number coincidences

It is appropriate to conclude this theoretical section with a brief discussion of famous numerical coincidences in cosmology (Sciama 1971). First, however we anticipate the results for the size parameters for the curvature-cosmology universe that come from section 6.2. These are shown in Table 3 where the N_H is the density divided by the mass of a hydrogen atom.

The first large number coincidence is the ratio of the radius of the universe to the classical electron radius (R/r_0). The result is 9.49×10^{40} which is to be compared with the ratio of the electrostatic force to that of the gravitation force between and electron and a proton. This is 4.3×10^{38} which being about 200 times smaller than R/r_0 shows that it is hardly a coincidence and although interesting probably has little physical significance.

Table 3: Size of curvature-cosmology universe

Quantity	Value	SI units
Radius, R	8.67 Gpc	2.68×10^{26} m
Volume, V	1.74×10^{31} pc^3	3.78×10^{80} m^3
Density, n_H	1.34 m^{-3}	2.24×10^{-27} kg.m^{-3}
Mass, M	6.22×10^{53} kg	6.32×10^{53} kg
$N_H = M n_H$	5.06×10^{80}	5.06×10^{80}

Sciama (1953, and 1971) investigated the use of Mach's principle and the role of inertia in general relativity. By direct analogy to Maxwell's equations, he derived *for rectilinear motion a combination of Newton's laws of motion and of gravitation, with the inertial frame determined by Mach's principle* (his italics). In effect, there is an acceleration term added to Newton's gravitational equation. The consequence is that the total energy (inertial plus gravitational) of a particle at rest in the universe is zero. He further assumed that matter receding with a velocity greater than that of light makes no contribution. The equivalent distance in curvature-cosmology is the radius, R. The implication of his theory is that

$$\frac{2\pi G\rho R^2}{c^2} \approx 1. \qquad (91)$$

Now using equation (68) we get the actual value for the left hand side in equation (91) to be ¾ and this value does not depend on the size of the universe. The closeness of this value to unity suggests that Sciama's ideas are worthy of further investigation.

6 Cosmological tests

This chapter compares the predictions of curvature-cosmology with observations. It must be pointed out that, particularly in the case of cosmology, there are no such things as 'pure' observations. In astronomy, observations are difficult and expensive. They are sometimes carried out specifically or implicitly on the basis of testing or demonstrating a particular cosmological theory. Even observations which were not carried out for the purpose of testing or supporting a specific theory may be analysed, interpreted and reported in such a way as to incorporate assumptions regarding a particular theoretical framework. These hidden assumptions need to be made explicit and to be removed if the observations are to be compared with predictions from a different theory. As far as possible, effects of Big-Bang cosmology have been explicitly removed from the observations discussed in this chapter. Nevertheless, it is likely that subtle effects that were not apparent or could not be removed from the data remain as contaminants.

6.1 Analytic methods

In the following sections, the predictions of curvature-cosmology are in many instances compared with observations by presenting the predicted and observed values as a function of redshift or some other variable in such a way that if curvature-cosmology is supported, the points should scatter about a horizontal straight line. That is the observations are transformed so that the expected distribution is a linear function of the independent variable. This form of display has two advantages. The first is that a qualitative impression of goodness of fit is provided since it is easy to see discrepancies from a straight line. The second advantage is that, to the first order, most deviations from curvature-cosmology can be approximated by a straight line with a slope that is not horizontal. Thus, the slope of the fitted line and its uncertainty provide a simple one-parameter estimate of goodness of fit. In most cases, the straight-line fitting procedure used is based on a statistical method that allows for uncertainties in both coordinates. Because this method is not widely known, it is briefly described as follows.

Traditional least squares regression provides two lines, the regression of y on x and the regression of x on y. In general, these regression lines do not have reciprocal slopes. If the data includes values for the uncertainties in both coordinates, the traditional minimisation of χ^2 methods would ignore the uncertainty in the independent variable. A better solution is obtained by the subroutine fitexy provided by Press et al. (1992) in their book Numerical Recipes, which also gives the basic references. This solution uses both sets of uncertainties in a completely symmetric manner that has a single solution regardless of which variable is deemed to be the independent. The method for estimating a and b in $y=a+bx$ is to minimise the χ^2 function defined by

$$\chi^2 = \sum_i \frac{(y_i - a - bx_i)^2}{\sigma_i^2 + b^2 \varepsilon_i^2}, \qquad (92)$$

where σ_i is the uncertainty in y_i and ε_i is the uncertainty in x_i. It is apparent that this equation has basic symmetry in that interchanging x and y will give the reciprocal of the slope b. Although it has no analytic solution, the minimisation can be done numerically. The uncertainties in the fitted parameters provided by fitexy (and quoted throughout this chapter) are a function of the uncertainties provided in the original data. The value of the fitted χ^2 is used only as a measure of the goodness of fit.

An important problem that arises in the use of observations is in how to handle anomalous data points. It is imperative that the omission of rogue data points does not in any way invalidate the comparison with theory by introducing bias in favour of (or against) the theory. However, in many cases the inclusion of these rogue points would seriously influence the value of χ^2 used to evaluate the comparison. The procedure adopted is conservative in that observations are rejected only if they have a very large χ^2 (with 1 DoF) against a smooth curve that passes through adjacent points. In other words, the rejection of a data point depends only on a comparison with its neighbours and not on assumptions about a theoretical model. Clearly, this procedure can be applied only when there are a large number of points, the great majority of which appear to lie on a relatively smooth curve.

6.1.1 Selection effects and bias

Most astronomical observations, in particular extra-galactic ones, have serious problems of interpretation due to strong selection effects. The major selection effect arises because most surveys of astronomical objects are limited to objects with flux density greater than some lower limit. The fainter ones cannot be distinguished above the observational noise. Usually this lower limit is determined by the telescope and this selection effect is often referred to as Malmquist bias (Malmquist 1920). In practice, the value of this lower limit is determined by a combination of observations of known objects combined with a detailed analysis of the telescope. Ideally, a full probability distribution of the detection probability would be available. In a review of the various forms of Malmquist bias Hendry, Simmonds & Newsam (1993) discuss several statistical approaches to overcoming the problem.

A typical problem that illustrates selection bias is that of determining the spatial frequency distribution of galaxies or quasars as a function of their absolute magnitude. The Malmquist bias means that fainter galaxies can be seen only at nearby distances. Large distances and hence large volumes are needed to achieve sufficient numbers of bright galaxies. In order to illustrate the effects of Malmquist bias in the measurement of galaxy and quasar density distribution Figure 3 shows the results of a Monte-Carlo simulation. For the simulation a Gaussian distribution of quasars was taken with a peak magnitude of -23 mag and a HPHM width of one mag (see equation (130)). This distribution is shown by the solid line. Next, the expected distribution for each of the three redshift ranges was estimated using the accessible volume analysis described in section 7.2.1 and they are displayed as points in Figure 3.

It is apparent that the omission of faint quasars moves the distribution to brighter quasars as the redshift region increases. Thus, any analysis for evolution that used this method would give erroneous results. There is a smaller but significant effect that exclusion of higher redshift quasars pushes the measured density distribution to fainter quasars. An analysis with the complete redshift range has an excellent fit to the assumed Gaussian distribution.

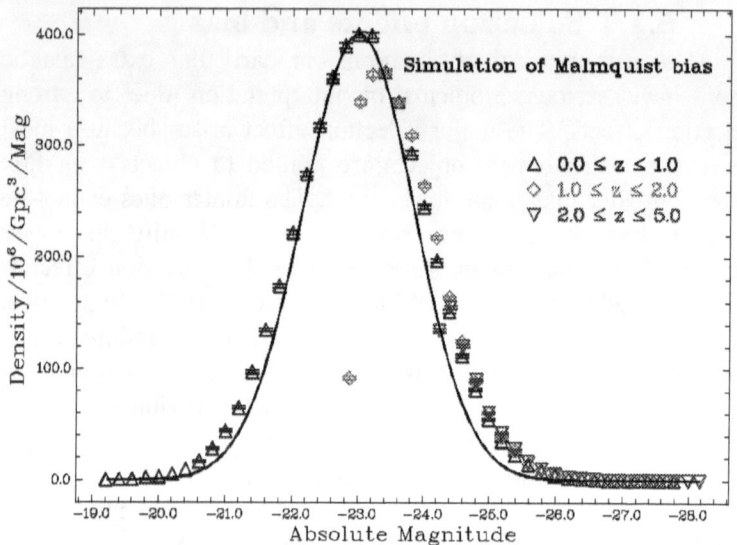

Figure 3: Simulation to show Malmquist bias in the measurement of the density distribution of galaxies and quasars.

In addition, distant galaxies and quasars have redshifts that significantly change the part of the spectrum that is sampled by the filters used. Furthermore, since galaxies have a wide variety of spectral types and it may be difficult to determine the spectral type of the fainter galaxies. Although the use of K-corrections and more recently the use of a large number of filters have gone a long way towards overcoming these difficulties, selection effects still produce major problems.

In using observations to evaluate a particular cosmological model, it is essential that each step in the analysis of the observations is examined and processed in terms of that model. Many astronomical investigations consist of identifying and determining the characteristics of a type of object, such as a galaxy, from nearby examples and then investigating their properties at much larger distances. While it is obvious that cosmological models are important in analysing the distant objects it is also true that the cosmological model may effect the nearby observations and produce a bias in the calibration characteristics. This can happen even though the differences in distance measures and linear sizes due to the different cosmological models are negligible.

In curvature-cosmology, the observed redshift of an object depends on the density and distribution of the gas between the object and us. This redshift is subject to the vagaries of the gas clouds. This is especially important for nearby objects: over large distances, the vagaries should average out. In Big-Bang cosmology in contrast, there is a precise relationship between distance and redshift: the Hubble law. In practice, this means that in curvature-cosmology redshift is an unreliable measure of distance for nearby galaxies. At best the use of redshifts to provide distances will only increase the uncertainty of the absolute magnitude (or linear size) measurements. At worst, there may be correlations that introduce a bias in the estimates. Thus, the fact that the Hubble relationship in Big-Bang cosmology is caused in a way that differs from that in curvature-cosmology means that differences in estimated values may be produced even over distances where the cosmological differences in distance measures are negligible.

Another example of possible bias occurs in the observations of spectral lines. If these lines are emitted or absorbed by large gas clouds the width of the lines is increased by curvature-redshift due to the density of the gas in the clouds. Many of the very wide absorption lines may be due to curvature-redshift. If these spectral lines are to be used to obtain astrophysical properties of the clouds, it is apparent that these measurements may be biased or even incorrect. A case in point is the use of relative strengths of spectral lines to measure the temperature of the gas. If the widths have a significant contribution from curvature-redshift then the temperature measurement could be grossly in error.

6.2 X-ray background

The first observations to be examined are those for the X-ray background. Since Giacconi et al. (1962) observed the X-ray background there have been many suggestions made to explain its characteristics. Although much of the unresolved X-ray emission comes from active galaxies, there is a part of the spectrum between about 10 keV and 1 MeV that is not adequately explained by emission from discrete sources.

In Big-Bang cosmology for the intermediate X-ray range of about 10 keV to 300 keV, the production of X-rays in hot cosmic plasma through the process of bremsstrahlung has been suggested by Hoyle (1963), Gould & Burbidge (1963), Field & Henry (1964) and Cowsik & Kobetich (1972). In a review of the spectrum of the X-ray background radiation Holt (1992) concludes that the measured spectra of discrete sources are not consistent with the observations in the intermediate energy range but there is a remarkable fit to a 40 keV (4.6×10^8 K) bremsstrahlung spectrum from a diffuse hot gas. However, in an expanding universe most of the X-rays are produced at redshifts of $z \geq 3$ where the density is large enough to scatter the cosmic microwave background radiation (CMBR). This scattering known as the Sunyaev-Zel'dovich effect (see section 6.4) makes a distinctive change in the spectrum of the CMBR. This predicted change for the densities required has not been observed and this is one of the main reasons why the bremsstrahlung model is rejected in Big-Bang cosmology.

In curvature-cosmology, the basic component of the universe is plasma with a very high temperature (section 5.2.2) and with low enough density to avoid the Sunyaev-Zel'dovich effect. The background X-ray emission is produced in this plasma by the process of free-free emission (bremsstrahlung). The observations of the background X-ray emission are analysed in order to measure the density and temperature of the plasma. In curvature-cosmology, this density is the only free parameter and it determines the size of the universe and the value of the Hubble constant. In addition, the temperature of the plasma determined from the X-ray measurements can be compared with the predicted value from curvature-cosmology of 2.56×10^9 K (section 5.2.2).

The first step is to calculate the expected X-ray emission from high temperature plasma in thermal equilibrium. Here the dominant mechanism is bremsstrahlung radiation from electron-ion and electron-electron collisions. With a temperature T and emission into the frequency range v to $v+dv$ the volume emissivity per steradian can be written (Nozawa, Itoh & Kohyama 1998) as

$$j_v(v)dv = \left(\frac{16}{3}\right)\left(\frac{\pi}{6}\right)^{1/2} r_0^3 m_e c^2 \left(\frac{m_e c^2}{kT}\right)^{1/2} g(v,T)$$
$$\times \exp\left(-\frac{hv}{kT}\right) N_e \sum Z_i^2 N_i dv \quad (93)$$

where $g(v,T)$ is the Gaunt factor, N_e is the electron density, N_i is the ion density and r_0 is the classical electron radius and the other symbols have their usual significance. In SI $j_v(v)$ has the units of W.m^{-3}.Hz^{-1}.

As it stands this equation does not include the electron-electron contribution. Nozawa, Itoh & Kohyama (1998) and Itoh, et al (2000) have done accurate calculations for many light elements. Based on their calculations Professor Naoki Itoh (http://www.ph.sophia.ac.jp/) provides a subroutine to calculate the Gaunt factor that is accurate for temperatures greater than 3×10^8 K that is used here. Let the average density be expressed as the number of hydrogen atoms per unit volume ($N=\rho/m_H$ m^{-3}). Then it is convenient to define $n_e = N_e/N$ and

$$n_i = \sum N_i Z_i^2 / N, \quad (94)$$

where the sum is over all species present. Because of the very high temperature, we can assume that all atoms are completely ionised. Thus, equation (93) including the Gaunt factor provides the production rate of X-ray photons as a function of the plasma temperature and density.

The next step is to compute the expected intensity at an X-ray detector. Consider an X-ray photon that is produced at a distance $R\chi$ from the detector. During its travel to the detector, it will have many curvature-redshift interactions. Although the photon is destroyed in each interaction, there is a secondary photon produced that has the same direction but with a slightly reduced energy. It is convenient to consider this sequence of photons as a single particle and to refer to it as a primary photon. The important result is that the number of these primary photons is conserved. Therefore, we need the production distribution of the number of photons per unit energy interval. The number of photons emitted per unit volume per unit time in the energy interval ε to $\varepsilon+d\varepsilon$ is given by

$$j_n(\varepsilon)d\varepsilon = \frac{j_v(v)}{\varepsilon}hdv, \qquad (95)$$

where $\varepsilon=hv$ and h is Plank's constant and $j_v(v)$ is the energy distribution per unit frequency interval. Now consider the contribution to the number of X-rays observed by a detector with unit area. Because the universe is static, the area at a distance $R\chi$ from the source is the same as the area at a distance $R\chi$ from the detector. Since there is conservation of these photons, the number coming from a shell at radius $R\chi$ per unit time and per steradian within the energy interval ε to $\varepsilon+d\varepsilon$ is

$$\frac{dn(r)}{dt}d\varepsilon = j_n(\varepsilon)d\varepsilon R d\chi. \qquad (96)$$

Next, we integrate the photon rate per unit area and per steradian from each shell where the emission energy is ε and the received energy is ε_0 to get

$$I_n(\varepsilon_0)d\varepsilon_0 = R\int_0^{\chi_m} j_n(\varepsilon)d\varepsilon d\chi, \qquad (97)$$

where $\varepsilon=(1+z)\varepsilon_0$ and it is assumed that the flux is uniform over the 4π steradians. Furthermore, it is useful to use equation (72) to change the independent coordinate to the redshift parameter z. Then using equation (95) and equation (69) we get

$$I_v(v_0)dv_0 = \frac{c}{H}\int_0^{z_m} \frac{j_v(v)}{1+z}dzdv_0, \qquad (98)$$

where H is the Hubble constant (equation (63)) and the change in bandwidth factor dv/dv_0, cancels the $(1+z)$ factor that comes from the change in variable from $d\chi$ to dz but there is another divisor of $(1+z)$ that accounts for the energy lost by each photon. Thus the energy flux per unit area, per unit energy interval, per unit frequency and per solid angle is given by equation (99) where Plank's constant is included to change the differential from frequency to energy. The z_m limit of 8.2 comes from the limit of $\chi \leq \pi$ in equation (71).

The density N is obtained by fitting equation (99) to the observed data as a function of the temperature T, and then extracting N from the normalisation constant. The X-ray data used is tabulated in Table 4. It consists of the background X-ray data cited in the literature and assessed as being the latest or more accurate results.

Table 4: List of background X-ray data used.

Name	Instrument	Reference
Gruber	HEAO 1 A-4	Gruber et al.1999
Kinzer	HEAO 1 MED	Kinzer et al.1997
Dennis	OSO-5	Dennis et al.1973
Mazets	Kosmos 541	Mazets et al.1975
Mandrou	Balloon	Mandrou et al.1979
Trombka	Apollo 16, 17	Trombka et al.1977
Horstman	Rocket	Horstman-Morretti et al.1973
Fukada	Rocket	Fukada et al.1975

$$I_\nu(\nu_0) = \left(\frac{16}{3}\right)\left(\frac{\pi}{6}\right)^{1/2} \frac{r_0^3 m_e c^3}{h} \left(\frac{3}{16\pi G M_H}\right)^{1/2} \left(\frac{mc^2}{kT}\right)^{1/2}$$

$$\times n_e n_i N^{3/2} \int_0^{z_m} \frac{g\left((1+z)\nu_0, T\right)}{1+z} \cdot \exp\left(-\frac{h(1+z)\nu_0}{kT}\right) dz$$

$$= \frac{2.3385 \times 10^3 \text{ keV}}{\text{keV.m}^2.\text{s.sr}} \left(\frac{mc^2}{kT}\right)^{1/2} n_e n_i N^{3/2} \times \quad (99)$$

$$\varepsilon_0 \int_0^{z_m} \frac{g\left((1+z)\nu_0, T\right)}{1+z} \cdot \exp\left(-\frac{h(1+z)\nu_0}{kT}\right) dz$$

Preliminary analysis showed that there were some discrepant data points that are listed in Table 5 in order of exclusion.

Table 5: Background X-ray data: rejected points

Source	Energy/keV	Flux density / keV/(keV.cm^2.s.sr)	χ^2 (1 DoF)
Gruber	98.8	0.230±0.012	108.6
Gruber	119.6	0.216±0.022	65.2
Fukada	110.5	0.219±0.011	66.6
Gruber	152.6	0.140±0.022	50.9
Fukada	179.8	0.110±0.005	41.5
Gruber	63.9	0.484±0.034	25.1

Very hard X-rays cannot be produced even by this hot plasma and are presumably due to discrete sources (Holt 1992). To allow for this component a power law was added to the bremsstrahlung spectra and its parameters were obtained by a χ^2 fit to data points above 550 keV. Since the bremsstrahlung spectrum was fitted to the original data points between the energy-limits of 10 KeV to 300 keV the power law fit for the high-energy points has no effect on the bremsstrahlung fit. Since bremsstrahlung is very sensitive to the presence of heavy elements, results are presented for four different abundances of hydrogen, helium, and 'metals'. The 'metals', which is a descriptor for all the other elements, were simulated by an element with Z=8.4, $<Z^2>$=75.3 and A=17.25.

Table 6: Abundances for four models.

Model	%H	%He	%metals	N_e/N	$\sum N_i Z_i^2 / N$
A	100.0	0.0	0.0	1.000	1.000
B	92.17	8.5	0.0	0.875	1.002
C	92.06	7.82	0.12	0.868	1.061
D	91.91	7.82	0.28	0.860	1.135

These values were derived from the abundances given by Allen (1976). The details of the four different abundances are shown in Table 6 where the percentages are by number and the last two columns show the relative number of electrons and average value of Z^2 per unit hydrogen mass. Thus the models are A: pure hydrogen, B: hydrogen with 8.5% helium, C: normal abundance and D: similar to C but with enhanced 'metals'.

Table 7: Fitted parameters for four abundance models.

Model		N	T_9	χ^2 (74 DoF)
A	Pure H	1.36±0.02	2.62±0.04	167.6
B	H plus He	1.34±0.01	2.62±0.04	167.6
C	H, He, and metals	1.02±0.01	2.61±0.04	168.5
D	Enhanced metals	0.77±0.01	2.61±0.04	169.1

The results of the fit of the data to these models is given in Table 7 where the symbol T_9 is the temperature in units of 10^9 K and the errors are the fitted uncertainties (1σ).

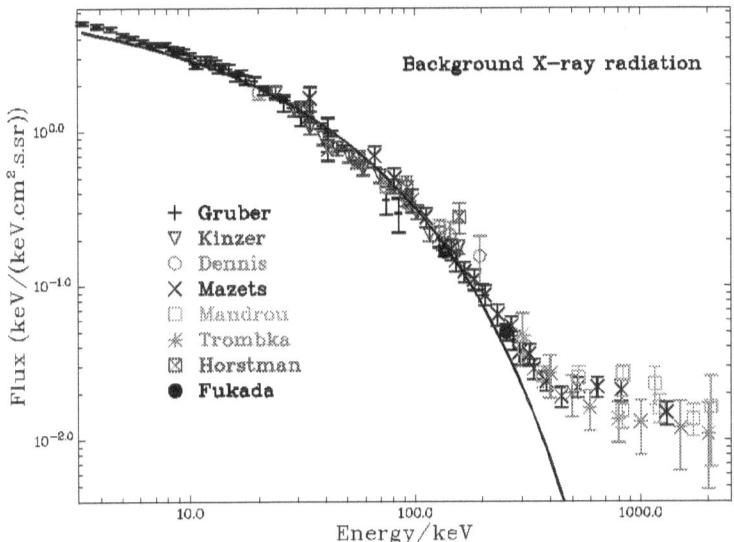

Figure 4: Background X-ray spectrum. See Table 4 for list of observations. Line is best fit from 10 keV to 300keV.

Figure 4 shows the flux density for the fitted curve for model B and for the observations as a function of energy. In order to clearly display the goodness of fit and avoid the clutter in Figure 4 the logarithm (base 10) of the ratio of the observed flux density to the predicted flux density (model B) is shown in Figure 5.

Most of the X-ray flux below 10 keV and part of the flux just above 10 keV is emission from discrete sources. The deviation from the curve at energies above about 300 keV arises from X-rays coming from discrete sources that have been modelled by a power law spectrum. In the intermediate region where bremsstrahlung should dominate, there are clear signs of some minor systematic errors. In addition, there appears to be some variation between the data sets. It is not clear whether the discrepancy between the observed points and the predicted flux densities is due to an inadequate theory, inadequate X-ray emission model, or systematic errors in the observations. After all the measurements are very difficult and come from a wide range of rocket and satellite experiments. In particular, the recent HEAO results (Kinzer et al. 1997) differ from earlier results reported by Marshall et al. (1980).

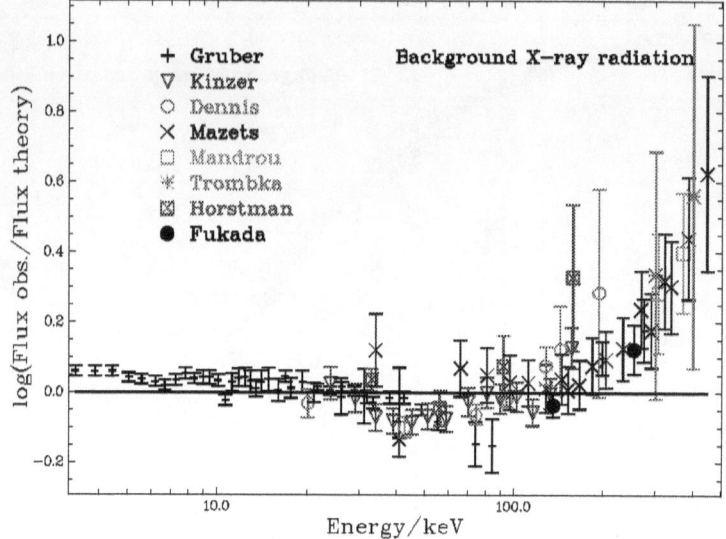

Figure 5: Background X-ray spectrum. Same as Figure 4 but showing log(data/theory)

It is apparent from Table 7 that although the measured temperature is relatively insensitive to the assumed abundance the density estimate is strongly dependent. This is because bremsstrahlung depends mainly on the number and to a lesser extent the type of charged particles whereas the density also depends on the number of neutrons in each nucleus.

Table 8: Fitted parameters for model B.

Quantity	Symbol	Value
Mean density	N	1.34 ± 0.01 m^{-3}
Electron density	n_e	1.18 ± 0.01 m^{-3}
Electron temperature	T_e	$(2.62 \pm 0.04) \times 10^9$ K
Power law value at 1 Mev	a	0.019 ± 0.001
Exponent	b	-0.54 ± 0.20

The power law fit parameters are the same for all the models and are shown in Table 8 for model B. This model was chosen because it uses the standard abundances that might be expected in the plasma and it has a relatively good fit to the observations. The quoted errors are the formal uncertainties of the fit. There are certainly larger, unknown systematic errors.

For the measured density of $N=1.34\pm0.01$ m^{-3} the calculated value of the Hubble constant is $H=(48.0\pm0.4)$ km.s^{-1}.Mpc^{-1} (from equation (63)). Further properties of the universe based on this density are shown in Table 3.

That said, this bremsstrahlung model for the background X-ray emission within a curvature-cosmology provides a good fit to the relevant observations. A crucial test of curvature-cosmology is that it predicts a temperature of 2.56×10^9 K for the cosmic plasma. The temperature estimated from fitting the X-ray data is $(2.62\pm0.04)\times10^9$ K. There is remarkable agreement between these values. It should be emphasised that the predicted temperature is a pure prediction from the theory without any dependence on observations. This agreement and the good fit to the observations gives strong support to curvature-cosmology.

6.3 Cosmic Microwave Radiation

Cosmic microwave background radiation (CMBR) was one of the most significant discoveries in cosmology in the 20th century. It has a black body spectrum to a high accuracy with a temperature of 2.725 ± 0.002 K (Mather et al. 1999). As well as the large dipole moment due to our motion, there is fine scale structure in the spatial distribution, some of which is associated with galactic clusters. These observations are so precise and so well verified that it is essential that any cosmological model should be able to explain them.

In curvature-cosmology, the CMBR comes from the curvature-redshift process acting on the high-energy electrons in the cosmic plasma (section 3.6). Examination of equation (31) shows that even for very high temperature plasma the emission from electrons will dominate that from other ions. The energy loss occurs when an electron that has been excited by the passage through curved spacetime interacts with a photon or charged particle and loses its excitation energy. Ideally, the theoretical model would provide the number distribution of secondary photons as a function of their energy. This distribution would then be combined with the distribution of electron energies to obtain the production spectrum of low-energy secondary photons from the plasma. The final step would be to integrate this production spectrum over all distances allowing for the geometry and curvature-redshift. The result would be the spectrum of photons that would be observed at any point in space. A demonstration of such a distribution was suggested in section 2.2 for the secondary photons produced by curvature-redshift applied to photons. However, uncertainties in its derivation do not provide the confidence needed to proceed with the complicated procedure involved in calculating an observed photon spectrum. Instead, arguments are given as to why the spectrum should be similar to that for a black body. Then based on observational evidence that the spectrum is very close to a black body spectrum we assume that the CMBR has a black body spectrum and then we estimate the temperature of the CMBR from the temperature and density of the cosmic plasma that were found in section 6.2. The CMBR photons also lose energy by curvature-redshift. The assumption of a black body spectrum enables us to equate the energy loss by the electrons to the energy loss by the CMBR photons and then to use Stefan's law to determine the temperature of the CMBR. After all the concept of temperature for the CMBR is only meaningful if it is in thermal equilibrium and thus has a black body spectrum.

The first point to note is that if there are no frequency constraints then the blackbody spectrum is the maximum entropy solution for a fixed energy density and will naturally occur if there is a mechanism for transferring photons between energy levels (Longair 1984) and all the energy levels are freely available. Usually this is done by the absorption and emission processes at the surface of a blackbody, hence the name. This brings up the problem of how the excited electrons produce the CMBR photons. Since parity conservation prevents a free electron from emitting dipole radiation, there must be an interaction with a third particle. A quick calculation shows that Thompson (Compton) scattering with the existing CMBR photons is too infrequent. The only other suitable interaction is Rutherford scattering off other electrons and ions. Since its last gravitational interaction, the electron will have become excited and have an excess of energy due to its passage through curved spacetime. This energy is transferred to secondary photons. Since the universe is equivalent to a very large box, the available photon energy levels are extremely close together and all will be available. The result is that we expect the production energy-spectrum of the low-energy CMBR photons to be a single humped distribution with a shape similar to a black body spectrum that can be described by a single parameter, the effective temperature.

The balance between the energy loss by the X-ray electrons and the energy loss by the CMBR photons implies that there is an overall conservation of energy with the photon energy loss being returned to the electrons. Since the secondary photons produced by curvature-redshift interaction of the CMBR photons have frequencies well below the plasma frequency (of about 10 Hz), their energy must go into plasma waves, which are dissipated with their energy going to heat the plasma. These processes are not driven by temperature differences so that there is no change in entropy.

For equilibrium, the energy gained by these photons must equal the energy lost by the photons. Since the dominant energy loss by photons in the cosmic space is via curvature-redshift, we can equate the two loss rates to determine the average energy of these photons. For electrons, or indeed any non-zero rest mass particle, the energy loss rate is given by equation (31). Using the metric of the cosmological model the energy loss rate for an electron is

$$\frac{d\varepsilon}{dt} = H\left[\beta^3(\gamma^2 - 1/2)^{1/2}(\gamma - 1)\right]m_e c^2, \qquad (100)$$

where to prevent confusion with the symbol for temperature the electron's kinetic energy is denoted by $\varepsilon = (\gamma-1)m_e c^2$ and the an extra factor of β comes from conversion of distance rate to time rate. The next step is to average this energy loss over the distribution of electron energies. Since the electrons are relativistic, the appropriate distribution is Jüttner distribution, which is (de Groot, Leeuwen & van Weert 1980)

$$n(p)d^3p = \frac{d^3p}{h^3}\exp\left(-\frac{\gamma mc}{kT_e}\right). \qquad (101)$$

With a change of variable to γ it becomes

$$n(\gamma)d\gamma \propto \gamma(\gamma^2-1)^{1/2} \exp\left(-\frac{\gamma m_e c^2}{kT_e}\right)d\gamma. \qquad (102)$$

Then integrating equation (100) over all the electron energies we get

$$\frac{d\varepsilon}{dt} = Hn_e m_e c^2 f(T_e), \qquad (103)$$

where n_e is the density of the electrons and $f(T_e)$ is average of the γ terms defined by

$$f(T_e) = \frac{\int_1^\infty \left[\left(\gamma^2-\frac{1}{2}\right)^{1/2}\beta^3(\gamma-1)\right]\gamma\left(\gamma^2-1\right)^{1/2}\exp\left(-\frac{\gamma m_e c^2}{kT_e}\right)d\gamma}{\int_1^\infty \gamma\left(\gamma^2-1\right)^{1/2}\exp\left(-\frac{\gamma m_e c^2}{kT_e}\right)d\gamma}. \qquad (104)$$

Although the Jüttner distribution can be integrated analytically in terms of modified Bessel functions, it is just as easy to evaluate both integrals numerically. Table 9 shows some values for the function $f(T_e)$ as a function of the electron temperature T_e.

Table 9: Some values for function f(T).

$T_e/10^9$	$f(T_e)$	$T_e/10^9$	$f(T_e)$	$T_e/10^9$	$f(T_e)$
1.2	0.138	1.8	0.443	2.4	0.967
1.3	0.175	1.9	0.515	2.5	1.076
1.4	0.217	2.0	0.592	2.6	1.193
1.5	0.265	2.1	0.676	2.7	1.316
1.6	0.318	2.2	0.767	2.8	1.445
1.7	0.378	2.3	0.863	2.9	1.582

The next step is to calculate the energy loss rate for the CMBR photons. If the CMBR photons are the result of curvature-redshift acting on the cosmic electrons and the system is in equilibrium these two loss rates should be equal.

If the CMBR photons were all produced at a fixed distance at a redshift z and if the production spectrum was a black body spectrum with a temperature T, then we would observe a spectrum with a temperature $T_0=T/(1+z)$. Thus, we expect the observed temperature to be less than the production temperature. However, the photons have been produced over a large range of distances so

Cosmic Microwave Radiation

that there must be an integration of the production spectrum over all distances.

Observations of the CMBR spectrum have shown that it is very close to a black body spectrum (Mather et al. 1990). If we assume a black body spectrum then the energy density of the CMBR photons near us must be the same as that for a uniform black body radiation with the same temperature. However, because the universe is homogeneous, the energy density must be the same everywhere. For equilibrium, the rate of energy loss by the high-energy electrons must equal the rate of energy loss by the CMBR photons. Then using equation (103) we get

$$\frac{4\sigma}{c}T_p^4 H = n_e m_e c^2 f(T_e) H, \text{ hence}$$

$$T_p^4 = 108.2161 n_e f(T_e)$$
(105)

where σ is the Stefan-Boltzmann constant and, not surprisingly, the Hubble constant cancels. Then from Table 8 we get n_e=1.18 and for a temperature of $(2.62\pm0.04)\times10^9$ K the calculated value of the function $f(T_e)$ is 1.215. These numbers result in a predicted temperature for the CMBR of 3.53 K. For the theoretical temperature of 2.56×10^9 K the predicted temperature of the CMBR is 3.48K.

Observations of the CMBR spectrum were obtained from the FIRAS instrument on the Cobe satellite by Mather et al. (1990). They measured the temperature of the CMBR to be 2.725 K. This temperature is confirmed by the observations of Roth & Meyer (1995) who measured a temperature of 2.729 (+0.023, -0.031) K using cyanogen excitation in diffuse interstellar clouds. The agreement within about 23% of the measured CMBR temperature with the predicted value is a strong endorsement of curvature-cosmology.

6.3.1 CMBR temperature at large redshifts

The temperature of the CMBR has been measured at large redshifts using two different methods. The first method measures the column density ratio of the absorption lines originating from the fundamental and first excited states of carbon (Ge, Bechtold & Black, 1995, Lima, Silva & Viegas 2000, and Srianand, Petitjean & Ledoux 2000). These lines are seen in the Lyman–alpha forest that is observed in the spectra from a bright quasar. The temperature estimate is based on the relative strengths of these spectral lines. For these measurements to be valid, it is essential that the line widths and column densities are well understood. However, as argued in section 6.11 curvature-redshift changes the widths of spectral lines and thus makes the usual interpretation of the spectra highly suspect (also see section 6.1.1). Thus, until this interpretation is fully understood the context of curvature-cosmology, CMBR temperature results from this method cannot be trusted.

The second method uses the Sunyaev-Zel'dovich effect acting on the CMBR by the gas in clusters of galaxies (Battistelli et al. 2002). By using multiple frequencies, it is possible to minimise the effects due to properties of the clusters on the result. However, as discussed in section 6.4, the method is flawed in curvature-cosmology because the CMBR has a different cause from that in Big-Bang cosmology. Thus, these results cannot be taken as showing a dependence of the temperature of the CMBR on redshift until the complete mechanism is understood in the context of curvature-cosmology.

6.4 The Sunyaev-Zel'dovich effect

The Sunyaev-Zel'dovich effect (Sunyaev & Zel'dovich 1970, Peebles 1993) is the effect of Thompson scattering of background radiation by free electrons in the intervening medium. The technique depends on knowing the spectrum of the background source and then measuring the changes in the spectrum due to the intervening plasma. In particular, it is the scattering in both angle and frequency of the cosmic microwave background radiation (CMBR) by electrons in the cosmic plasma. The effect is often characterised by the dimensionless Compton y-parameter, which for a distance x through non-relativistic thermal plasma with an electron density of N_e has the value

$$y = \frac{kT_e}{m_e c^2} \sigma_T N_e x = 3.44 \times 10^{-16} N_e T_e x \quad (x \text{ in Mpc}), \tag{106}$$

where σ_T is the Thompson cross-section. An object at redshift z is at the distance

$$x = \frac{\sqrt{2}c}{H} = 7.103 N^{-1/2} \log(1+z) \text{ Gpc}. \tag{107}$$

Hence, using $T_e = 2.75 \times 10^9$ K, $N_e = 1.18$ m^{-3} we get

$$y = 9.2 \times 10^{-6} \log(1+z). \tag{108}$$

Using the CMBR as a source the Sunyaev-Zel'dovich effect has been observed and Mather et al. (1990) report an observed upper limit of $y = 0.001$, and more recently Fixsen et al. (1996) report $y = 1.5 \times 10^{-5}$. However, this analysis is done for a Big-Bang cosmology where the CMBR arises in the early universe and has a black body spectrum. Although we assume that in curvature-cosmology, the CMBR has a black body spectrum it is predominantly produced in the region where z is less than one. Although the measurements are consistent with production from a black body with z less than 5, the full story must await a proper analysis of where the CMBR is produced in curvature-cosmology and for a re-analysis of the observations carried out for this situation.

6.5 Type 1a Supernovae

The type 1a supernovae are believed to be the result of an explosion of a white dwarf that has been steadily acquiring matter from a close companion. When the mass exceeds the Chandrasekar limit, the white dwarf explodes producing a very bright supernova whose light curve shows a rapid rise over several weeks, then an equally rapid fall followed by a much slower decrease over several hundred days. The type 1a supernovae are distinguished from other types of supernovae by the absence of hydrogen lines and the occurrence of strong silicon lines in their spectra near the maximum. Although the theoretical modelling is poor, there is much empirical evidence that they all have remarkably similar light curves, both in absolute magnitude and in their time scales. This has led to a considerable effort to use them as cosmological probes. Since they have been observed out to redshifts with z greater than one they can be used to verify the cosmological time dilation that is predicted by expanding cosmologies. Figure 6 shows a plot of the widths defined by equation (110), of the light curves as a function of redshift (the

SCP data used is described below), where the nearby supernovae are shown with open circles and rejected supernovae are shown with filled circles. Indeed the results, which show excellent quantitative agreement with the predicted time dilation, have been hailed as one of the strongest pieces of evidence for an expanding cosmology. As currently understood, these results present a challenge to any static cosmology.

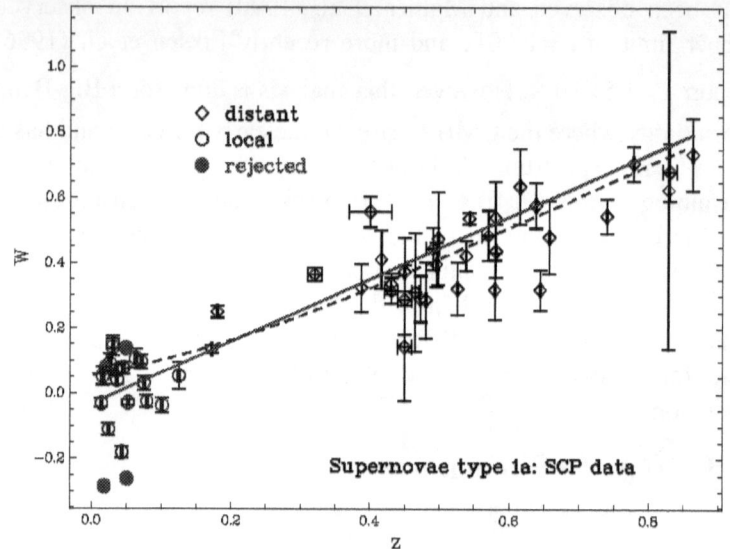

Figure 6: Width of supernovae type 1a verses redshift for SCP data. Solid line is best linear fit. Dashed line is curvature-cosmology prediction.

To address this challenge it is necessary to explore in detail how these observations were obtained and to examine the assumptions underlying their analysis. For each supernova, there are three relevant variables: the absolute magnitude at the peak of the light curve, the width of the light curve and the redshift. A cosmological model is needed to compute the absolute magnitude from the apparent magnitude. It will be shown by comparing Big-Bang analysis with curvature-cosmology analysis that the cosmological model has a significant influence on both the gathering and the interpretation of existing observations of type 1a supernovae.

We have used data from the Supernova Cosmology Project (SCP) where the original width data comes from Goldhaber et al. (2001) and the magnitudes come from Perlmutter et al. (1999). Their 42 new supernovae are in addition to 18 local supernovae that came from the Calàn/Tololo Supernovae Survey. The original apparent magnitudes have been corrected for the K-correction, extinction within our Galaxy and extinction in the host-galaxy. They analysed all of these 60 supernovae by the same method of fitting a standard light-curve template to the observations of each supernova. A subsequent paper from the group (Knop et al. 2003) adds 11 new supernovae and provides a better analysis for some of the earlier supernovae. To avoid bias only 54 supernovae that were deemed to be valid by Knop et al. (2003) are used. The z errors were taken from Perlmutter et al. (1999) and the magnitude errors come from the column labelled 'd' in tables 3, 4 and 5 in Knop et al. (2003). To remove one of the assumptions of Big-Bang cosmology, the stretch corrections (see equation (109)) were removed from the magnitudes. The advantage of this data is that all the supernovae parameters have been obtained by a consistent method. Accurate K-corrections were provided by Kim, Goobar & Perlmutter (1996). From these 54 candidates four supernovae were removed because of very poor fits to the width verses redshift plot. Table 10 gives a list of these supernovae and their χ^2 values for a fit to the other 50 supernovae. Since these are all relatively nearby (low values of z), their omission has little cosmological impact.

Table 10: Omitted supernovae

Supernovae	Redshift	χ^2
1990af	0.050	363.8
1992bc	0.020	183.9
1992bo	0.018	1347.5
1995ac	0.049	99.6

Although this is a large project with complex observational techniques and required extensive data analysis, the results that are relevant here are the redshift z, apparent magnitude m, and relative width w of the light curve for each supernova. The widths were

78 Curvature Cosmology

derived from the stretch factors s, provided by Knop et al. (2003) by the equation

$$w = (1+z)s. \tag{109}$$

For consistency with the magnitudes and to make it easier to interpret straight lines as power laws, we have defined a new variable W by

$$W = 2.5\log(w). \tag{110}$$

Except for a multiplying constant (with a value of 1.086) W is essentially identical to w for the range of widths considered here. Figure 6 shows a plot of the width parameter W verses the redshift parameter z where the local Calàn/Tololo supernovae are shown as open circles and the rejected supernovae are shown as filled circles. The straight line is the best fit and has the slope 0.956 ± 0.022 with a goodness of fit $\chi^2 = 509.4$ (48 DoF). What is surprising is the poor fit. After all, if the slope is due to cosmic expansion, the redshifts and predicted widths are accurately known. Although Perlmutter et al. (1999) provided detailed error budgets for the magnitudes and widths this poor fit suggests that there is additional intrinsic dispersion in the widths that has not been included in the original errors. That said, the significant positive slope of the line is not, on the face of it, consistent with curvature-cosmology.

6.5.1 Analysis with curvature-cosmology

Curvature-cosmology interprets the apparent increase of light-curve width with redshift as shown in Figure 6 as being due to selection of the supernovae. This interpretation depends on the results that follow. First, we examine the dependence of absolute magnitude on width and find a strong relationship that is opposite to that which is currently accepted (Phillips 1993, Hamuy et al. 1996, and Supernovae Cosmology Project). On the other hand, it is consistent with a physical model in which the type 1a supernovae can have variable widths and absolute magnitudes subject to the constraint that the total energy output is a constant. Second, we show that for the local supernovae used here and using curvature-cosmology the dependence found by Phillips (1993) is not statistically significant. Third, and most importantly from the historic perspective, we show that there is no dependence of absolute magnitude on width if the Big-Bang luminosity-distance relationship is used to calculate absolute magnitudes.

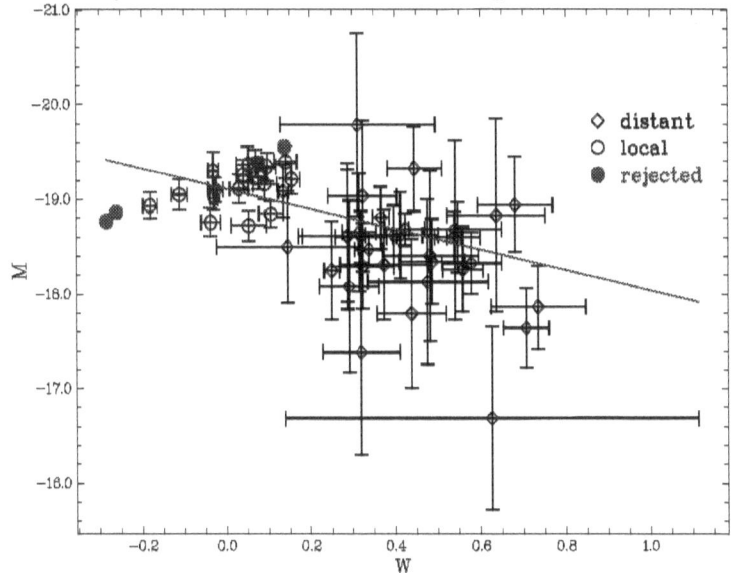

Figure 7: Absolute magnitudes of type 1a supernovae verses widths of their light curves in curvature-cosmology. SCP data.

For curvature-cosmology, the luminosity-distance relation is given by equation (81) that has only one free parameter, the Hubble constant and since this is an additive constant, its actual value is not important in this context. However, in order to have realistic magnitudes for the plots, the analysis is done with h=0.7 (i.e. H=70 km.s^{-1}Mpc^{-1}). A plot of absolute magnitude verses width is shown in Figure 7 where the nearby supernovae are shown with open circles and the rejected supernovae are shown filled circles. The best-fit straight line has a slope of 1.07±0.18 and χ^2=210.2 (48 DoF) (Note that the vertical axis is in the numerically reverse order that puts the brightest supernovae towards the top.) Supernovae that are brighter have narrower widths, or the weaker are wider.

The magnitude-width relationship is thus an intrinsic property of type 1a supernovae. Then, in curvature-cosmology, the width can be used to correct the observed absolute magnitude to a reference value that makes an excellent 'standard candle'. Alternatively, the absolute magnitude can be used to correct the observed width to the reference width.

To assess whether this dependence of absolute magnitude on width shown in Figure 7 is realistic it is necessary to understand what it physically means. The current model for type 1a supernovae (Hillebrandt & Niemeyer 2000) is that a white dwarf in a binary system gradually accretes matter from its companion. Eventually it exceeds the limits of its stability – the Chandrasekhar limit – and there is a thermonuclear incineration of the white dwarf. Because this is a well-defined limit, it is expected that each supernova has the same mass and therefore the same energy output. The light curve has a rise time of about 20 days followed by a fall of about 20 days and then a long tail that is most likely due to the decay of ^{56}Ni. The widths are measured in the light coming from the expanding shells before the radioactive decay dominates. Thus the widths are a function of the structure and opacity of the initial explosion and do not depend on the radioactive decay. The fitted slope shows that the luminosity – width relationship is

$$L \propto w^{-1.07\pm0.18}. \tag{111}$$

However, the total energy observed is the integral of the luminosity over time. Assuming that each supernova has the same shaped light curve we get the total energy to be $E \propto Lw \propto w^{-0.07\pm0.18}$. Thus, the total energy is essentially constant. We may reasonably conclude that this relationship is consistent with the known physics of the explosion.

It is well known (Phillips 1993, Hamuy et al.1996, and Supernovae Cosmology Project) that there is a correlation between the width of the supernovae light curve and its absolute magnitude. However, the dependence that they found was obtained from the local supernovae and has the opposite sense to the slope shown by Figure 7. In order to illustrate this dependence, Figure 8 shows a plot of the widths verses absolute magnitudes for the local supernovae with, as before; the rejected supernovae are shown by filled circles. The straight line has a slope -1.16±0.47 with χ^2=24.9 (15 DoF). As can be seen the dependence is not statistically significant. Although the inclusion of the excluded supernovae makes this slope significant (slope -1.31±0.28 with

$\chi^2=31.3$ (19 DoF)), we have no explanation as to why it has the opposite slope to that shown in Figure 7. We can only surmise that because most of these supernovae were found by a wide variety of methods there may some unknown selection effects present (see section 6.1.1). Although for nearby supernovae the distance measure is essentially, independent of the cosmological model it does require accurate distances and if they are derived from redshifts then in curvature-cosmology many of the absolute

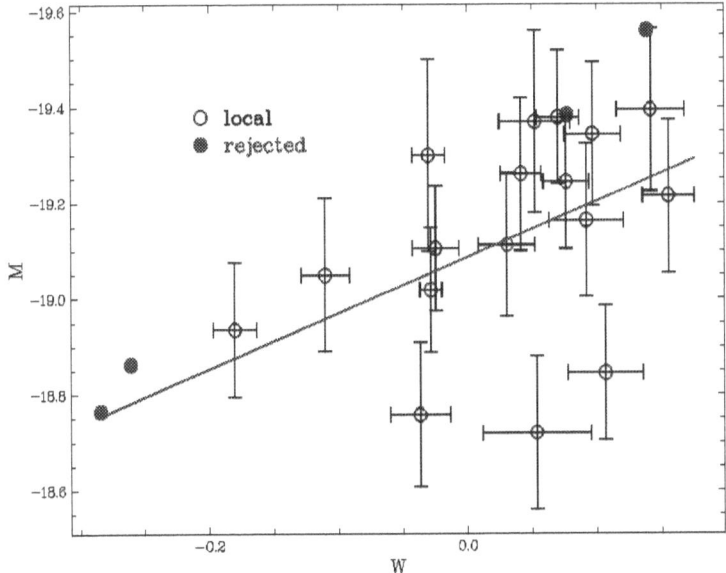

Figure 8: Absolute magnitude verses width for local type 1a supernovae. SCP data.

magnitudes may be in error.

Note that this slope of -1.31 shows that the more luminous supernovae have wider light curves and that the luminosity-width relationship is

$$L \propto w^{1.31\pm0.28}.$$ (112)

However, the total energy observed is the integral of the luminosity over time. Then as before we assume that each supernova has the same shaped light curve, we get the total energy to be $E \propto Lw \propto w^{2.31\pm0.28}$. This shows that the energy output is a strong function of the width. Physically this is less plausible that the constant energy output described by equation (111). Since both cosmologies have essentially the same luminosity-distance relationship for these nearby supernovae, their analysis is only relevant as to whether there is a large intrinsic relationship between absolute magnitudes and widths. However, it should be noted that in Big-Bang cosmology any corrections implied by equation (112) would use the stretch factor and not the observed width.

We also note that the range of these widths of the light curves from local supernovae is about 0.5 and the range in magnitudes is about one. This is consistent with results from Richardson et al. (2002) who have done a comparative study of the absolute magnitude distributions of supernovae. For type 1a supernovae they find that after eliminating doubtful outliers, the distribution is approximately Gaussian with a standard deviation of 0.56 mag.

If this dependence of absolute magnitude on width is, so strong it is pertinent to ask why this dependence was not discovered earlier. The answer is that in Big-Bang cosmology the luminosity-distance relationship implicitly includes the width dependence. We can show this by repeating the analysis as shown in Figure 7, of absolute magnitude verses width but using the Big-Bang luminosity-distance relationship. The fitted straight line has a slope -0.04±0.16 with χ^2=45.6 (48 DoF). As expected, the insignificant slope means that no dependence of absolute magnitude on width is observed for distant supernovae in Big-Bang cosmology. This is consistent with the premise that the supernovae were all selected to have essentially the same absolute magnitude (using the Big Bang model). In addition the standard deviation of their absolute magnitudes is very small. For the 50 supernovae (SCP data) the rms is 0.22 mag which is smaller than the value of 0.56 mag quoted by Richardson et al (2002).

The result that the power law shown by equation (111) has a slope consistent with one means is that the property that is common to all type 1a supernovae is their total energy of emission. The emission may occur as a narrow light curve with a high maximum luminosity or a wide light curve with a lower maximum luminosity. Therefore, we can use the width-luminosity relationship to correct the widths to a standard value or use the widths to correct the absolute magnitudes to a 'standard candle'.

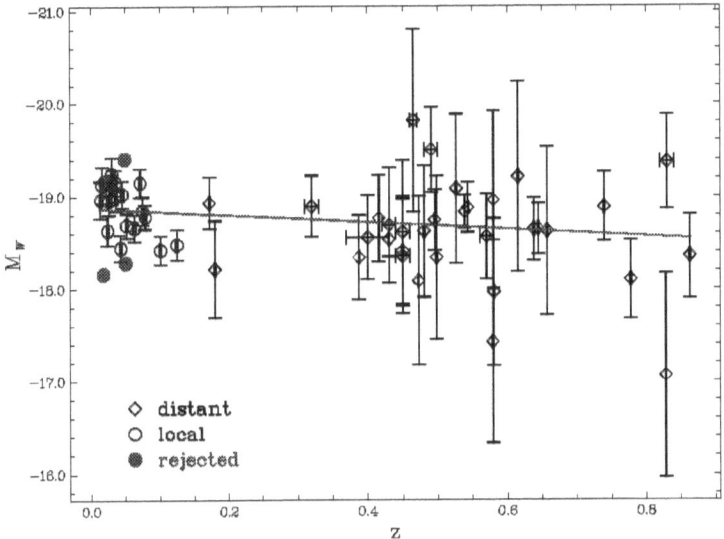

Figure 9: Absolute magnitudes corrected for light curve widths verses redshift. SCP data.

The other important test is to see if the absolute magnitudes (corrected for width) are independent of redshift. Figure 9 shows a plot of corrected absolute magnitudes verses redshift. The straight line has a slope of -0.38 ± 0.16 with $\chi^2=91.5$ (48 DoF). This slope is not statistically different from being horizontal. Thus, the absolute magnitude corrected for width is independent of redshift. This also shows that the luminosity-distance relationship in curvature-cosmology is in good agreement with supernovae data. Thus, there is a simple relationship between width and absolute magnitude that is in agreement with the physically realistic concept that the total energy output of type 1a supernovae is a constant.

The next step is to show that there is a selection that depends on absolute magnitude that determines which supernovae are observed. It is this selection that produces the apparent dependence of width on redshift.

6.5.2 Supernovae: selection effects

Since curvature-cosmology requires that the strong width verses redshift dependence shown in Figure 6 is not due to the expansion of the universe but that it is due to selection effects it is necessary to have a reasonable model of the selection effects for the cosmology to be credible. A linear fit of the width corrected for absolute magnitude verses redshift has a slope of -0.35±0.15 with χ^2=65.9 (48 DoF) and this shows that the corrected width is essentially independent of redshift. However, this does not completely explain Figure 6. If there were no selection effects the distribution of observed widths at any redshift should cover the complete distribution of widths. That is without selection Figure 6 should show a scatter plot without any systematic variation with redshift. Clearly, there is a systematic variation of observed widths with redshift that needs to be explained.

We have shown that supernovae have the necessary range in magnitudes (and widths) then what selection effects could produce the apparent dependence of width on redshift? The technique for the supernovae observations is a two-stage process (Perlmutter & Schmidt 2003, Strolger et al. 2004). The first stage is to conduct repeated observations of many target galaxies to look for the occurrence of supernovae. Having found a possible candidate the second stage is to conduct extensive observations of magnitude and spectra to identify the type of supernova and to measure its light curve. This second stage is extremely expensive of resources and it is essential to be able to determine quickly the type of the supernova so that the maximum yield of type 1a supernovae is achieved. Since current investigators assume that the type 1a supernovae have essentially a fixed absolute magnitude (with possible corrections for the stretch factor), one of the criteria is to reject any candidate whose predicted absolute magnitude (at maximum light) is outside a rather narrow range. The selection effect to be considered is one where all candidates whose absolute magnitude is ΔM magnitudes brighter than the canonical value are rejected. The essential point is that the absolute magnitudes are calculated using Big-Bang cosmology and hence the selection of candidates is dependent on the luminosity-distance relationship given by Big-Bang cosmology. Another assumption is that the intrinsic distribute of absolute magnitudes is a Gaussian with a standard deviation of σ.

The first step in the selection model is to select a hypothetical supernova with a fixed absolute magnitude and fixed width, then to calculate the apparent magnitude using curvature-cosmology. The second step was to compute an absolute magnitude using Big-Bang cosmology (using equation (85) with $\Omega_M=0.3$). Next, the average absolute magnitude ΔM is computed from the Gaussian distribution with an upper cut-off of $-\Delta M/\sigma$, where $-\Delta M$ is the difference between the original and the final absolute magnitudes. Then with the assumption of fixed energy $W(z)$ is proportional to ΔM and $W(z)$ can be plotted as a function of z. Figure 6 shows a plot off the observations together with the theoretical curve (dashed line) of $W(z)$ for the case when $\Delta M = 0.5\sigma$. What this means is that if the supernovae had an intrinsic distribution in absolute magnitude with a standard deviation of σ and if all candidates that appeared to be too bright by about 0.5σ magnitudes are rejected then the dependence is explained. Clearly, this magnitude cut-off is smaller than what would be used. However in a comprehensive description of the selection procedure Strolger et al. (2004) state:

Best fits required consistency in the light curve shape and peak color (to within magnitude limits) and in peak luminosity in that the derived absolute magnitude in the rest-frame B band had to be consistent with observed distribution of absolute B-band magnitudes shown in Richardson et al. (2002).

Thus, there are additional selection criteria that depend on the colour and the width of the light curve, which would strengthen the selection effect. Since the time dilation of the supernova light-curve widths was expected and was deemed to have been proven by earlier observations, it was reasonable to include the width of the light-curve as a criterion in the selection process. After all the prime purpose of these investigations was to determine the structure and evolution of the universe by studying the magnitudes of type 1a supernovae as a function of redshift. This simple analysis does not prove that selection effects were responsible for the light curve width dependency on redshift but it does show that this dependency could well be explained by selection effects.

6.5.3 Type 1a supernovae: magnitudes

It has been shown that a type 1a supernova makes an excellent 'standard candle' if its light-curve width is used to correct its absolute magnitude. Now we investigate the luminosity-distance relationship in curvature-cosmology by seeing if the corrected absolute magnitudes of type 1a supernovae are independent of redshift.

Although Figure 9 shows that the corrected magnitudes for the SCP supernovae are independent of redshift the redshift range only goes to z=0.863. Riess et al. (2004) have provided a larger sample of supernovae cover a larger redshift range and that have been analysed in a consistent manner. There is considerable overlap with the supernovae in the SCP most of the supernovae do not have width measurements. However, they have the advantage of greater numbers and a larger redshift range. However, since all these supernovae were subject to the same selection process, we can use the strong width verses redshift relationship shown in Figure 6 to define proxy widths. The proxy width is defined to be $w_p=1+z$. Riess et al. (2004) have done the corrections for the stretch factors in a way that makes it difficult to remove their effects on magnitude: these stretch corrections should not cause a bias but could slightly increase the scatter.

The plot of absolute magnitudes, corrected for the proxy width, from Riess et al. (2004) using their 'gold' supernovae are shown in Figure 10 as a function of redshift. Following Riess et al. (2005) the absolute magnitude reference value was changed from -19.44 mag to -19.17 mag.

Examination of Figure 10 shows that the large numbers of nearby supernovae appear to dominate the zero point. As already discussed (see section 6.1.1), all the redshifts of galaxies include curvature-redshifts due to the gas halo of both our Galaxy and the target galaxy. If they are in a cluster, this can be augmented by curvature-redshifts in the inter-galactic gas. The effect is that estimating distance from their redshift over-estimates the distance and hence makes absolute magnitudes appear to brighter than they should. The effect is most severe for nearby galaxies and negligible for the distant ones. In order to minimise the problem a straight line is fitted to the same data is plotted in Figure 10 (dashed line) but omitting those supernovae that have redshifts with z less than 0.1. The slope is 0.061±0.088 with χ^2=57.0 (62 DoF) and with an intercept of -19.08±0.06. Thus, the absolute magnitudes are essentially constant which shows that curvature-cosmology can explain the absolute magnitudes of supernovae after they are corrected for the widths of the light curves.

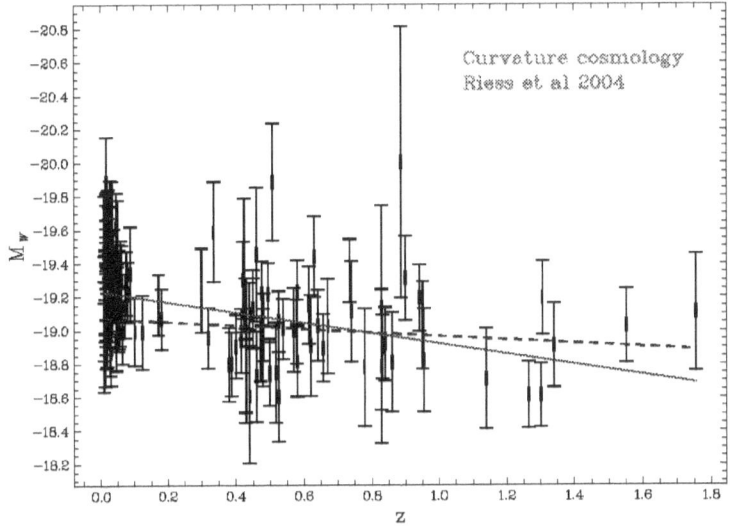

Figure 10: Absolute magnitudes for supernovae type 1a for data from Riess et al (2004), corrected for proxy widths. Dashed line omits supernovae with z<0.1.

Knop et al. (2003) found that the best fit to the supernovae type 1a magnitudes using Big-Bang cosmology gave the density parameter Ω_M=0.25, which implies that the cosmological-constant energy-density parameter Ω_Λ=0.75. These investigators agree with earlier suggestions that there has been acceleration in the universal expansion in recent times. There have been two explanations for this acceleration: 'dark energy' Turner (1999) and 'quintessence' Steinhardt & Caldwell (1998). Although there are theoretical arguments based on vacuum energy and basic theoretical problems with Big-Bang cosmology there is no other observational evidence for these concepts. However in curvature-cosmology this acceleration is meaningless and there is no evidence for anomalies in the data. Hence, there is no need for dark energy or quintessence.

6.5.4 Type 1a supernovae: SNLS data

Recently (Astier et al. 2005) have published some of the first results from the Supernovae Legacy Survey (SNLS). Although some of the local supernovae are common to those from the Supernovae Cosmology Project (SCP: Knop et al. 2003), all of them both local and distant have been analysed by a consistent method that has been designed to avoid any bias. The data from Astier et al. (2005) has been analysed in exactly the same method as above for the SCP supernovae. The first step was to reject three supernovae that had a very bad fit in the width verses redshift analysis. They are listed in Table 11 in order of rejection.

The next step was to examine the dependence of width on redshift. A fit for a straight line of the width as a function of redshift has the functional form

$$W = 0.879 \pm 0.002 + (1.213 \pm 0.005)z \qquad (113)$$

with χ^2=9336.8 (112 DoF). Again, we note the very large value of χ^2.

Table 11: Reject supernovae from Astier et al. (2005)

Name	Redshift	χ^2 (1 DoF)
SNLS-04D2gp	0.707	5959.1
1990af	0.050	2128.5
1999aw	0.039	1332.5

To assist comparison of the two sets of supernovae data Table 12 lists the slopes of straight-line fits for various combinations of parameters. The notation M_W means a width corrected for absolute magnitude by the equation

$$M_W = M_{origional} + 1.07W, \qquad (114)$$

which is applied to each supernova using its given width. The inverse relationship is used to get the corrected width W_M.

It is apparent that the residual χ^2 values for the SNLS data are very large, either due to under-estimates for the uncertainties (which is unlikely) or due to some unknown additional component. As a consequence, the quoted uncertainties in column

6 are almost certainly too small. Finally, M_{BB} is the absolute magnitude calculated with the Big-Bang luminosity-distance relationship described in section 6.5.1 but no width corrections.

Table 12: Slopes of fitted straight lines.

	Vert.	Horiz.	SCP Slope	SCPχ^2 48 DoF	SNLS Slope	SNLS χ^2 112 DoF
1	w	z	1.10±0.02	632.9	1.213±0.005	9336.8
2	W	z	0.96±0.02	509.4	1.035±0.005	6804.6
3	M	W	1.07±0.18	71.2	2.120±0.016	38616.9
4	W_M	z	0.35±0.15	65.9	0.267±0.009	47597.1
5	M_W	z	0.38±0.18	65.9	-0.286±0.010	47597.1
6	M_{BB}	z	0.04±0.16	45.6	-0.277±0.006	86771.8

As shown in rows 1 and 2 of Table 12 there is reasonable agreement between the SCP and SNLS data sets for the apparent width dependence on redshift. Of interest is that for both data sets a linear fit with W is better than a linear fit with w. However, row 3 shows that there is a large difference between them for the absolute-magnitude width dependence. We can only surmise that this is due to the different selection procedures. Rows 4 and 5 show that for SCP data that there is no dependence of the corrected width or the corrected absolute magnitude on redshift but there is a dependency for SNLS data. Finally row 6 shows that there appears to be strong dependence of absolute magnitude on width for SNLS data when it is computed using Big-Bang cosmology. This suggests that there is some aspect of the SNLS supernovae selection or analysis that makes a bias in the SNLS width estimates. This is why the SNLS slope in row 3 was disregarded in computing the corrected magnitudes and widths. Until this bias and the very large χ^2 values are resolved, the SNLS data must be treated with caution. Nevertheless, the SNLS type 1a supernovae results are not inconsistent with curvature-cosmology.

6.5.5 Supernovae analysis: summary

Since the arguments used to explain Figure 6 in curvature-cosmology are somewhat complex it may be helpful to re-visit them without the complexities of the analysis. In summary the main explanations for supernovae observations in curvature-cosmology are

- There is a strong intrinsic relationship for type 1a supernovae between absolute magnitude (at peak power) and the width of the light curve,
- The relationship is consistent with an energy output that is the same for all supernovae.
- The absolute magnitude can be used to correct the observed width to a standard width.
- Alternatively, the width can be used to correct the absolute magnitude to get a 'standard candle'.
- There is a sufficient distribution in intrinsic widths to produce the currently observed distribution of widths.
- The process of selecting supernovae to be observed has biased the selection in that only larger widths are selected at larger redshifts.
- The distribution of (corrected) absolute magnitudes is independent of redshift.

In Big-Bang cosmology the explanations are

- Any intrinsic relationship between absolute magnitude and widths is small and in the opposite sense to the one in curvature-cosmology.
- The widths have the expected time dilation as shown by Figure 6.
- There is no observed relationship between absolute magnitude and widths.
- The distribution of absolute magnitudes as a function of redshift is not constant but requires the universe to be accelerating.
- This acceleration is explained by 'dark energy' or 'quintessence'.

For both cosmologies, the supernova observations are self-consistent. The major criticism of curvature-cosmology arguments is the requirement that there is a biased selection process. Whereas the Big-Bang arguments can be criticised for having to introduce ad hoc physics such as 'dark energy'.

6.5.6 Supernovae analysis: conclusion

Both Big-Bang and curvature-cosmology models can explain the supernovae results. Big-Bang cosmology requires the ad hoc addition of dark energy (or quintessence) to explain the apparent acceleration of the expansion of the universe. On the other hand, given the physically plausible intrinsic spread of light-curve-widths, curvature-cosmology model needs selection effects to obtain the width verses redshift relation displayed in Figure 6. A simple model of possible selection effects shows that the idea is plausible but it needs more details of what selections were actually used for complete resolution. Provided the absolute magnitudes are corrected for width, curvature-cosmology shows that the distribution of absolute magnitudes of type 1a supernovae is independent of redshift. In curvature-cosmology, there is no anomaly that is equivalent to universal acceleration in Big-Bang cosmology. Given strong selection effects, the observations of type 1a supernovae show excellent support for curvature-cosmology.

6.6 Quasar variability in time

One of the major differences between a tired-light cosmology and an expanding universe cosmology is that any expanding universe cosmology predicts that time variations and clocks have the same dependence on redshift as does the frequency of radiation. Hawkins (2001, 2003) has analysed the variability of 407 quasars over an eighteen-year period. His data permitted the straightforward use of Fourier analysis to measure the time scale of the variability. He showed that there was no significant change in the time scale of the variability with increasing redshift. A χ^2 test showed that the power spectra of the two redshift bins ($0.5<z<1.7$ and $1.7<z<3.5$) were compatible at the 30% level. He considered and rejected various explanations including that the time scales of variations were shorter in bluer pass bands or that the variations were not intrinsic but were due to intervening processes such as gravitational micro-lensing. His conclusion was either that the quasars are not at cosmological distances or that the expanding universe cosmologies are incorrect in this prediction.

Curvature-cosmology on the other hand would predict just such results.

More recently, Vanden Berk et al. (2004) have analysed the variability of ~25,000 quasars from the Sloan Digital Sky Survey (see section 7.3). They derived a measure of the magnitude variability and analysed it as a function of time lag (i.e. the time scale of the variability), quasar luminosity, rest-frame wavelength, and redshift. In summary their results are: the variability increases with time lag (up to about 2yr), quasars that are more luminous are less variable, shorter wavelengths are more variable, and more distant quasars are somewhat more variable. Unfortunately, they did not investigate a direct dependence between time lag and redshift and it is important to note that they corrected the observed time lags to the rest-frame time lag. In curvature-cosmology, the time lags are independent of redshift. The complexity of the strong selection effects and the other dependencies mentioned make it difficult to use their data to resolve any possible dependency between time-lag and redshift. For example if the time lags are not corrected to the rest-frame then the variability may have the opposite sign and decrease with increasing time-lag.

To summarise, the sparse data on quasar variability strongly supports curvature-cosmology and is inconsistent with the notion that time scales vary with redshift.

6.7 The linear size of radio sources

One of the important tests of the geometry of a cosmological model is how the angular size of an object varies with redshift. Since the linear size of individual objects is unknown, the geometry is tested by measuring the angular size of a class of objects, computing their linear size, and then examining if some average of these linear sizes varies with redshift. If the computed linear sizes do not show a variation with redshift then we have some confidence that the geometry of curvature-cosmology is correct. For curvature-cosmology the relationship between the linear size, d, and the angular size θ is

$$d = R\sin\left(\frac{\ln(1+z)}{\sqrt{2}}\right)\theta, \qquad (115)$$

where the small angle approximation is used and z is the redshift parameter. Putting $R=10$ Gpc and measuring θ in arc seconds it is

$$d = 48.48 \sin\left(\frac{\ln(1+z)}{\sqrt{2}}\right) \theta'' \text{ kpc} . \qquad (116)$$

Because of their large size, double-lobed radio sources make suitable candidates for investigating the linear size of objects as a function of redshift. Gurvits, Kellerman & Frey (1999) have measured the size of 330 extended radio sources. They used 5GHz VLBI contour maps taken from the literature to measure the angular size. After selection for spectral index and flux density, they measured the distance between the core and the most distant component that had a peak brightness of at least 2% of the peak brightness of the core. They identified the optical sources as galaxies, BL Lac objects or quasars. The current analysis is restricted to 254 quasars that have a well-defined distance between the core and a 2% component. The linear size of these sources is shown in Figure 11 as a function of redshift.. The lower dashed

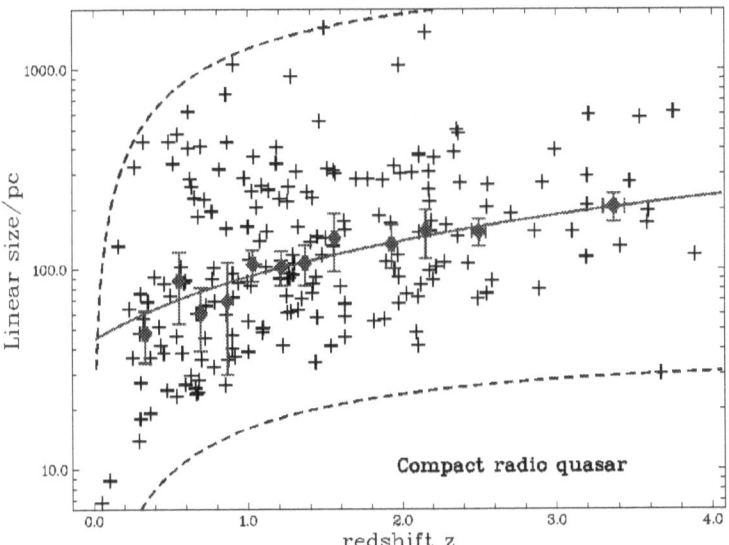

Figure 11: Linear size of compact radio quasars verses redshift. Data from Gurvits et al (1999).

line shows the locus of the source with the smallest angular size as if it had been seen at different redshifts. The upper dashed line is the similar locus for the source with the largest angular size. These two lines give an idea of the selection limits for their survey.

Because of the wide range of sizes, the mean is a poor measure of the linear size as it is too sensitive to outliers. The median is a better measure. The data was divided into bins with 17 sources in each bin (the 3 quasars with lowest redshifts were omitted) and the median was computed for each bin. The uncertainty in the median value was computed as the mean absolute difference from the median. These points are also plotted in Figure 11. A linear straight line fit to the 11 medians with $z>0.5$ gave a slope of 47.3 ± 10.9 kpc.z^{-1}. Note that because of the logarithmic scale the 'straight' line of best fit appears curved.

Except for low redshifts, where cosmological effects are minimal, the data are consistent with curvature-cosmology. Selection effects can easily explain the slight increase in linear size with redshift. As well as any intrinsic scatter, the apparent size is modified by projection effects. A double radio source whose components are close to the line of sight will not be resolved unless the angular separation of the components is greater than the survey limit. Clearly, this translates into an increasing cut-off in linear size as redshift increases. In addition, there is a bias (discussed below) due to the flux density limit of the survey.

Buchalter et al. (1998) carefully selected 103 double-lobed sources from the VLA FIRST survey and measured their angular sizes. Using quasar catalogues, they selected edge-brightened, double-lobed radio sources and measured the angular size directly from the FIRST radio maps.

These are Faranoff-Riley type II objects (Faranoff & Riley 1974) and exhibit radio-bright hot-spots near the outer edges of the lobes. A plot of their results is shown in Figure 12. The same analysis as above was done with these sources and the median values are shown in Figure 12. Note that these objects are about one order of magnitude larger than the core objects used by Gurvits, Kellerman & Frey (1999). In this case, the slope for $z>0.5$ was 0.39 ± 0.15 Mpc.z^{-1}.

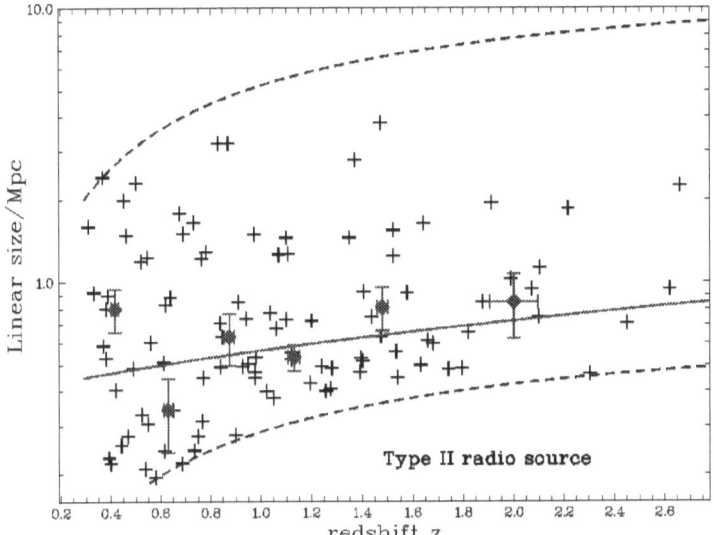

Figure 12: Linear size of Faronoff-Riley type II radio sources verses redshift. Data from Buchalter et al (1998).

There is clear evidence of selection bias in these linear sizes that is most evident at low redshifts. Given that this bias is strongest at low redshifts, the cosmological effects of this bias are minimal. Another bias in the selection procedure that may influence the results is that in both cases the samples are flux density limited. The effects of this bias are difficult to estimate but if larger structures correspond to larger flux densities, we would expect that a flux density cut-off would result in selection of larger structures at higher redshifts. This would appear as a slow increase in linear size with redshift that is seen in both plots.

For both compact radio sources with a typical linear size of about 100 kpc and for very large radio sources with a linear size of about 1 Mpc, the results show excellent agreement with curvature-cosmology.

6.8 Tolman surface brightness

In a series of papers Sandage and Lubin (Sandage & Lubin 2001, Lubin & Sandage, 2001a, 2001b, 2001c) have investigated the Tolman surface brightness test for elliptical and S0 galaxies. This test, suggested by Tolman (1934), relies on the observation that the surface brightness of objects does not depend on the geometry of the universe. For a uniform source, the quantity of light received per unit angular area is independent of distance. However, the quantity of light is sensitive to non-geometric effects, which make it an excellent test to distinguish between cosmologies. For expanding universe cosmologies the surface brightness is predicted to vary as $(1+z)^{-4}$, where one factor of $(1+z)$ comes from the decrease in energy of each photon due to the redshift, another factor comes from the decrease in rate of their arrival and two factors come from the apparent decrease in area due to aberration. In a static, tired-light, cosmology only the first factor is present. Thus to show that surface brightness observations are consistent with curvature-cosmology we must show that the surface brightness varies as $(1+z)^{-1}$.

The obvious candidates for surface brightness tests are elliptic and S0 galaxies for which projection effects are reduced. The major problem is that surface brightness measurements are intrinsically difficult due to the strong intensity gradients across their images. The observational difficulties are thoroughly discussed by Sandage & Lubin (2001) with the conclusion that the use of Petrosian metric radii helps solve many of the problems. Petrosian (1976) (and Djorgovski & Spinrad 1981, Sandage & Perelmuter 1990) showed that the ratio

$$\eta = 2.5\log\left(2\frac{d(\log(r))}{d(\log(L(r)))}\right), \tag{117}$$

where r is the angular radius and $L(r)$ is the total luminosity within that radius, is an experimental measure of a suitable radius. Note that the argument of the outer logarithm is a derivative. It is identical to the difference in magnitude between the mean surface brightness $<SB(r)>$, within the radius r and the surface brightness at r, namely $SB(r)$.

Thus, the aim is to measure the mean surface brightness for each galaxy at the same Petrosian radius. Because of the high correlation that exists for surface brightness as a function of different Petrosian radii, we need use only the data for a single Petrosian radius ($\eta=1.7$). Thus, the choice of Petrosian radii greatly diminishes the differences in surface brightness due to the luminosity distribution across the galaxies. However, there still is a dependence of the surface brightness on the size of the galaxy.

The purpose of the preliminary analysis is not only to determine the low z absolute luminosity but also to determine the surface brightness verses linear size relationship that can be used to correct for effects of size variation in distant galaxies. The data on the nearby galaxies used by Sandage & Lubin (2001) was taken from Postman & Lauer (1995) and consists of extensive data on the brightest cluster galaxies from 119 nearby Abell clusters. For consistency similar procedures to those used by Sandage & Lubin (2001) have been used here, except that we use $H=70$ km.s^{-1}Mpc^{-1}. The non-linear corrections given in their Table 3 and the 0.16 mag zero offset were applied to the surface brightness values given in their Table 1. The only difference was that curvature-cosmology has been used to compute the linear distance. In practice, this makes only a small difference for the redshifts involved. The higher z data comes from Lubin & Sandage (2001c). They made Hubble Space Telescope observations of galaxies in three clusters and measured their surface brightness and radii. The names and redshifts of these clusters are given in Table 13.

Table 13: Cluster names and redshifts

Name	1324+3011	1604+4304	1604+4321
Redshift	0.6565	0.8967	0.9243

Figure 13 shows a scatter plot of the logarithm of the linear radius (in pc) verses the absolute magnitude (Cape/Cousins R band) within a Petrosian radius of $\eta=1.7$. The magnitudes and surface brightness of the two clusters CL1324+3011 and CL1604+4304 were corrected from the I band to the R band by increasing them by 0.62 magnitudes (Sandage & Lubin 2001c). The data is shown twice, once at the top using Big-Bang cosmology and with an offset of +2. The second at the bottom uses curvature-cosmology.

It is apparent that the nearby galaxies are typically larger in size and luminosity than the cluster galaxies. Apart from the use of curvature-cosmology luminosity and distance relationship, the main reason for this difference is that the nearby galaxies are the brightest in their cluster and the distant ones are normal cluster members. It is also clear that the linear radius calibrations are needed for much smaller radii and the non-linear corrections (which have been applied) are important.

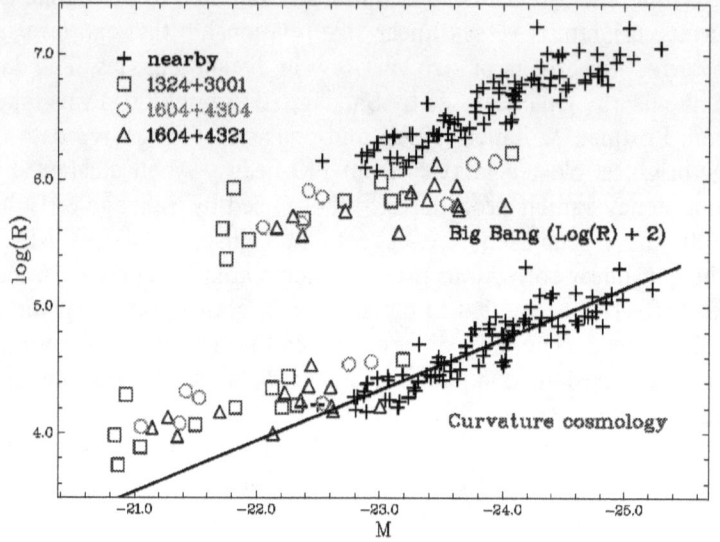

Figure 13: Log(R) verses absolute magnitude. R is the linear size in pc. Line is the best fit straight line to nearby galaxies using curvature-cosmology.

Of secondary interest is that if all the galaxies are included in a straight-line fitted to the local galaxies shown in Figure 13, the slope of -0.401±0.053 for $\log(R)$ verses M shows that the luminosity L, as a function of the linear radius R, can be approximated by a power law with an exponent 1.02±0.05. This is consistent with the luminosity being proportional to the radius.

The next step in the preliminary analysis is to correct the surface brightness for the absolute size of the galaxy. Figure 14 shows a plot of the surface brightness verses the logarithm of the linear radius. The straight line is the best fit for the data from the local galaxies and has the form

$$<SB> = (8.67 \pm 0.48) + (2.86 \pm 0.10)\log(R) \qquad (118)$$

These coefficients are similar to those found by Sandage & Lubin (2001). Because the surface brightness of the distant galaxies should vary with redshift, we do not expect their points to be close to the straight line. Since we are considering static, stationary cosmology in which there is no evolution as a function of redshift, we expect the surface brightness to vary as $(1+z)^{-1}$.

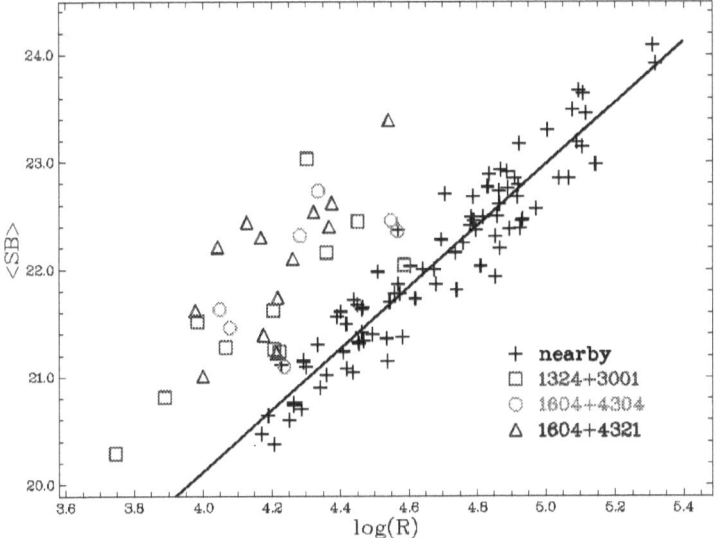

Figure 14: Surface brightness verse log(R). R is the linear radius in pc. The line is the best fit straight line to nearby galaxies.

Figure 15 shows a plot of the surface brightness with each value corrected for its linear radius by using equation (118) verses $2.5\log(1+z)$. This variable was used so that expected power laws for the surface brightness would have a straight-line dependence. The slope (fitted to all the data) is -1.81 ± 0.11. This may be compared with the slope found by Lubin & Sandage (2001c) of -1.41 ± 0.17. While neither result agrees with the expected exponent of -1.0, Figure 15 shows that the surface brightness data shows better agreement with curvature-cosmology than it does with Big-Bang cosmology. A major problem with this data is that distant galaxies and nearby galaxies are not drawn from similar populations. In particular, the data was for the brightest members of the nearby clusters but for all the members of the distant clusters.

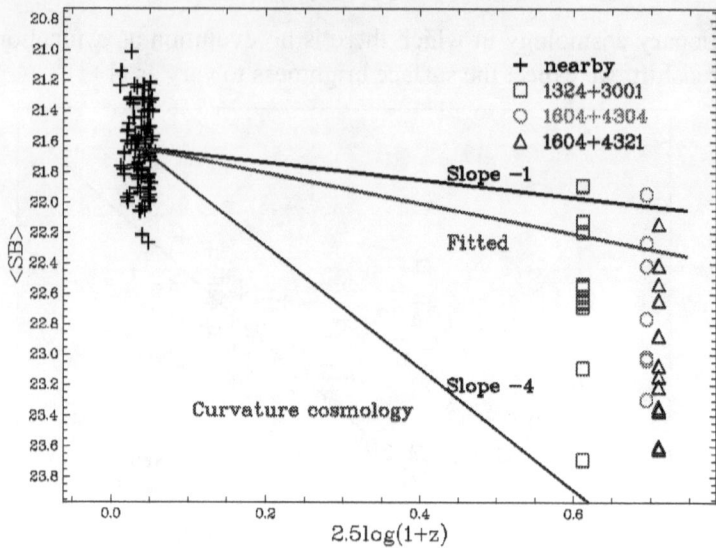

Figure 15: Surface brightness verses 2.5log(1+z).

We can partially overcome some of the selection problems by restricting the analysis to the nearby galaxies and only include the three brightest members of the distant clusters. Then the fitted slope is -0.58±0.24 that gives a strong hint that a more thorough selection and analysis would show a much better agreement with curvature-cosmology. Therefore, we conclude that the surface brightness data shows strong support for curvature-cosmology.

6.9 Clusters of Galaxies

Among the strongest observational evidence that led to the need to hypothesise the existence of dark matter in Big-Bang cosmology is that the mass derived from the use of the virial theorem and galactic velocities in a cluster of galaxies is very much larger than the mass estimated from galactic luminosities. The virial theorem is a very general and powerful statistical theorem that states that for a system of bodies in equilibrium, the average kinetic energy of the bodies is simply related to the average force on a body multiplied by a scale factor. If the galaxies in a cluster are gravitationally held together by their mutual gravitational attraction, it is found that the mass predicted by the virial theorem can be several orders of magnitude greater that that determined by luminosities. Even the mass of inter-galactic gas is not enough to overcome this imbalance. In Big-Bang cosmology dark matter has been introduced to make up for the shortfall of mass.

However, if curvature-redshift is valid then it is possible that the observed redshifts are not due to kinematic velocities but are curvature-redshifts produced by the inter-galactic gas. The purpose of this section is to show that curvature-redshift can explain the galactic velocities without requiring dark matter.

For simplicity, we will use the Coma cluster as a test bed. Not only is it very well studied, but it also has a high degree of symmetry and the presence of an inter-galactic gas cloud is known from X-ray observations. Watt et al. (1992) and Hughes (1989) have fitted the density of the gas cloud to an isothermal β-model with the form

$$\rho = \rho_0 \left(1 + \left(\frac{r}{r_e}\right)\right)^{-\alpha}, \qquad (119)$$

with a centre at 12h59m10s, 27°59′56″ (J2000) and with $r_e = 8'.8 \pm 0'.7$, $\alpha = 1.37 \pm 0.09$, $\rho_0 = 2.67 \pm 0.22\, h_{50}^2 \times 10^3\, m_H.m^{-3}$. The central density is obtained from the X-ray luminosity and has a strong dependence on the distance. Watt et al. (1992) assumed a Hubble constant of 50 km.s^{-1}.Mpc^{-1}. With a mean velocity of 6,853 km.s^{-1} (Colless & Dunn 1996) and with this Hubble constant, the distance to the Coma cluster is 137 Mpc. Recently Rood et al. (1993) using the Tully-Fisher relation to measure the distance modulus to the galaxies in the Coma cluster, to observe a value of 34.4±0.2 mag whereas Liu & Graham (1993) using infrared surface brightness fluctuations get 34.99±0.21 mag. The average is 34.7 mag that corresponds to a distance of 87.1 Mpc. This is consistent with the distance of 85.6 Mpc given by Freedman et al. (2001). Thus putting $h_{50}=1.57$ gives a corrected central gas density of $\rho_0=(6.61\pm0.54)\times10^{-3}\, m_H.m^{-3}$.

The galactic velocity data is taken from Beijersbergen & van der Hulst (2003) who provide information for 583 galaxies. The velocity centroid of the Coma cluster is 12^{h}59^{m}19^s, 27°52′2″ (J2000). They find that early-type galaxies (E+S0+E/S0) have a mean velocity of 9,926 km.s^{-1} and a rms (root mean square) velocity of 893 km.s⁻¹. Let us assume that all the galactic velocities are due to curvature-redshift. That is we assume that the actual velocities, the peculiar velocities, are negligible. Then the redshifts for the galaxies are calculated (in velocity units) using equation (11) by

$$v = \int_0^Z 51.691\sqrt{N(Z)}\, dZ \text{ km.s}^{-1}, \qquad (120)$$

where Z is the distance from the nearest edge of the Coma cluster to the galaxy measured in Mpc, and $N(Z)$ is the density of the inter-galactic gas cloud. The problem here is that we do not know Z distances. Nevertheless, we can still get a good estimate by assuming that the distribution in Z is statistically identical to that in X and in Y. In a Monte Carlo simulation, each galaxy was given a Z distance that was the same as the X (or Y) distance of one of the other galaxies in the sample chosen at random. For 50 trials, the computed dispersion was 554 km.s⁻¹ which can be compared with the measured dispersion of 893 km.s⁻¹ (for all 583 galaxies). Curvature-cosmology has predicted the observed dispersion of galactic velocities in the Coma cluster to within a factor of two. Considering that, this is a prediction of the cosmological model without fitting any parameters and ignoring all the complications of the structure both in the gas and galactic distributions the agreement is remarkable.

Since the distance to the Coma cluster is an important variable, the computed velocity dispersion from the Monte Carlo simulation for some different distances (all the other parameters are the same) is shown in Table 14. Thus, the redshift dispersion (in velocity units) is approximately a linear function of the Coma distance. This is not surprising since in this context the distance is mainly a scale factor.

Table 14: Coma velocity dispersions for some distances.

Distance/Mpc	50	87	100	150
Dispersion /km.s⁻¹	318	554	636	955

Beijersbergen & van der Hulst (2003) note that a better fit to the velocity distribution is provided by the sum of two Gaussian curves. Their best fit parameters for these two Gaussians are $v_1=7{,}501\pm187$ km.s⁻¹, with $\sigma_1=650\pm216$ km.s⁻¹ and $v_1=6{,}640\pm470$ km.s⁻¹, with $\sigma_1=1{,}004\pm120$ km.s⁻¹. This double structure is supported by Colless & Dunn (1996) who argue for an ongoing merger between two sub clusters centred in projection on the dominant galaxies NGC 4874 and NGC 4889. In addition, Briel, Henry & Böhringer (1992) found evidence for substructure in the X-ray emission and Finoguenov et al. (2004) and White, Briel & Henry (1993) have measured the X-ray luminosity of individual galaxies in the Coma cluster showing that the model for the gas used above is too simple. The net effect of this substructure is that the observed velocity dispersion would be different from that predicted by a simple symmetric model. Thus, it appears that substructure makes it very difficult to achieve a more accurate test of curvature-cosmology using the Coma cluster.

There is an important difference between curvature-cosmology and models that assume that the redshifts of the galaxies within a cluster are due to their velocities. Since the laws of celestial mechanics are symmetric in time, any galaxy could equally likely be going in the opposite direction. Thus a galaxy with a high relative velocity could be in the near side of the cluster or equally likely on the far side of the cluster. However, if the redshifts are determined by curvature-redshift then there will be a strong correlation in that the higher redshifts will come from galaxies on the far side of the cluster. A possible test is to see if the apparent magnitudes are a function of relative redshift. With a distance of 87.1 Mpc the required change in magnitude is about 0.025 mag.Mpc^{-1}. A simple regression between magnitude of Coma galaxies (each relative to its type average) and velocity did not show any significant dependence. Although this was disappointing, several factors can explain the null result. The first is the presence of substructure; the second is that the magnitudes for a given galactic type have a standard deviation of about one magnitude, which in itself is sufficient to wash out the predicted effect; and thirdly mistyping will produce erroneous magnitudes due to the different average velocities of different types. In support of the second factor we note that for 335 galaxies with known types and magnitudes, the standard deviation of the magnitude is 1.08 mag and if we assume that the variance of the Z distribution is equal to the average of the variances for the X and Y distributions then the expected standard deviation of the slope is

0.076 mag.Mpc^{-1}. Clearly, it is such larger than the expected result of 0.025 mag.Mpc^{-1}. It is expected that better measurements or new techniques of measuring differential distances will in the future make this a very important cosmological test.

Russel (2002) has measured the distances to many galaxies using a variety of methods and found an anomalous velocity distribution in the Virgo cluster. Although the Virgo cluster is highly irregular, it is a well-known X-ray object and thus has inter-galactic gas. A plot of the velocities of galaxies and distances provided by Russel (2002) is shown in Figure 16. Galaxies within a projected radius of less than 3 Mpc are shown in filled circles and those outside are shown as squares.

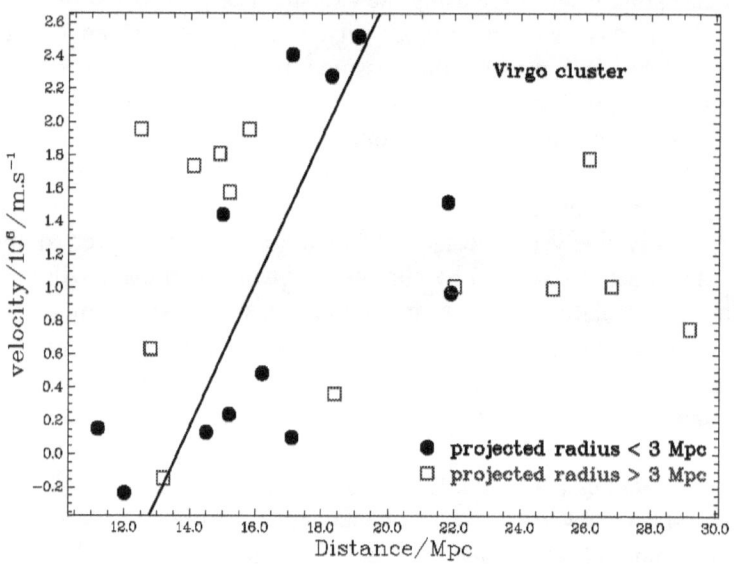

Figure 16: Redshifts (in velocity units) verses actual distance for some Virgo galaxies. Solid circles show those with projected radius less than 3Mpc.

Curvature-redshift model predicts that there should by a large non-linear increase in apparent velocity due to the density of the inter-galactic gas. The galaxies outside 3 Mpc are presumably outside the major part of the gas cloud and are not expected to show a significant redshift effect. The straight line is a weighted fit to the galaxies within 3 Mpc of the cluster centre. For the fit it was assumed that the distances had an uncertainty of 0.3 Mpc and that the 'velocities' had an uncertainty of 100 kms^{-1}.

The straight line has a slope of 436±152 km.s^{-1}.Mpc^{-1} (χ^2=10.2, 12 DoF) which is reasonable compared with the predicted Coma velocities considering that the effect is non-linear and is a strong function of the position of the line of sight relative to the centre of the gas cloud. Figure 16 shows very strong support for the occurrence of curvature-redshift in the Virgo cluster.

6.9.1 Related cluster properties

There is a long history (Wu & Xue 1999) of relationships between the X-ray luminosity and temperature of the intra-cluster gas and the velocity dispersion of the cluster galaxies. Although there are explanations based on gravitational potentials and the virial theorem, curvature-cosmology provides an alternative and simpler model. The X-ray luminosity L_X is a function of the temperature and is proportional to the square of the gas density. However, the apparent velocity dispersion σ, is proportional to the square root of the gas density. Thus with a very simple scaling model and constant temperature we would expect that $L_X \propto \sigma^4$. Plionis & Tovmassian (2004) looked at the X-ray luminosity of groups and found an exponent of 4.1±0.6. Although this agrees with the prediction, they had to make corrections for underestimation of the velocity dispersion of elongated groups when seen roughly orthogonal to the line of sight. It is a bias due to the small number of galaxies observed in the groups.

The analysis is complicated by the dependence of X-ray luminosity on temperature. In fact, Wu & Xue (1999) find that in separate analyses for temperature and dispersion

$$L_X \propto T^{2.72 \pm 0.05} \text{ and } L_X \propto \sigma^{5.24 \pm 0.29}. \tag{121}$$

Although the theoretical temperature dependence of luminosity on temperature is well known, we do not know whether the gas is isothermal, adiabatic or has some other relationship between pressure and temperature as a function of radius. Consequently, a more thorough analysis is not warranted until a better temperature-pressure relationship is known.

If curvature-redshift is valid then the redshift of the galaxies in the cluster will have been increased by the additional redshift due to the inter-galactic gas. Thus, they will have, on average, a larger redshift than an isolated galaxy at the same distance.

Table 15 shows the predicted (effective) velocity for a galaxy in the centre plane of the Coma cluster as a function of the projected radius. The second column is the velocity at that exact radius and the third column shows the average velocity of galaxies (uniformly spread in area) within that radius. This simulation also showed that the average velocity offset for the galaxies in the Coma cluster is 1206 kms^{-1} which means that the 'redshift' of the centre of the Coma cluster is 6926-1206=5720 kms^{-1}. This offset is important for calculating the Hubble constant, which from these figures is 5270/87.1=65.7 kms^{-1}.Mpc^{-1}.

Table 15: Velocity at, and average velocity within various projected radii in the Coma cluster (distance = 87.1Mpc).

Projected radius /Mpc	Velocity /km.s^{-1}	(Mean velocity < radius)/km.s^{-1}
0.0	2327.7	2327.7
0.5	1477.7	1764.8
1.0	1033.4	1342.5
1.5	803.3	1096.9
2.0	658.6	933.2
2.5	557.0	814.4
3.0	481.0	723.3
3.5	421.7	650.7
4.0	374.0	541.2
4.5	334.8	541.2
5.0	302.0	498.7

In addition, the redshift of objects seen through a cluster will be increased by curvature-redshift from the inter-galactic gas. Karoji, Nottale & Vigier (1976) claim to have seen this effect. They examined radio galaxies and classified them into region A if their light does not pass through a cluster and region B if their light passes through a cluster. They found no significant differences in magnitudes between the two regions but they did find a significant difference in the average redshifts that was consistent over the complete range. Their result is that radio galaxies seen through a cluster had an average extra redshift (in

Clusters of Galaxies 107

velocity units) of 2412±1327 km.s^{-1}. Overall the difference in the distance modulus was ΔDM=0.16±0.04, which is significant. Since the density and distribution of the gas in the clusters is unknown and the limiting radius of the cluster is not stated it is impossible to get an accurate prediction. Nevertheless, we note that for the Coma cluster with a radius of 2 Mpc the average extra redshift (from Table 15 with a factor of two) corresponds to 1866 km.s^{-1} showing that curvature-cosmology could explain the effect. In a different study, Nottale (1976) compared the magnitude of the brightest cluster galaxy with that in another cluster with similar redshift. He found that there was no significant difference in magnitudes between clusters but that the clusters with the largest number of galaxies had the higher redshift difference between the pairs. On average the redshift difference (in velocity units) was 292±85 km.s^{-1}. This can be explained by the expected correlation between number of galaxies and size and density of the inter-galactic gas.

In his review of voids in the distribution of galaxies, Rood (1988) quotes Mayall (1960) who observed a large void in the distribution of galaxies in front of the Coma cluster. This void has a magnitude of about 3000 kms^{-1}, which although somewhat larger, is not inconsistent with the expected value of about 1200 km.s^{-1}. In other words, the Coma cluster galaxies have an extra curvature-redshift due to the inter-galactic gas. However, the galaxies just outside the cluster nearer to us do not have this extra redshift and would appear to be closer to us. Hence, we see an apparent void in the redshift distribution in front of the Coma cluster.

A consequence of gas clouds and curvature-redshift is that the distribution of redshifts is similar to but not identical to the distribution of z distances. Galaxies that are behind a cloud will have a higher redshift than would be expected from a simple redshift distance relationship. Thus, we would expect to see anomalous voids and enhancements in the redshift distribution. This will be accentuated if the gas clouds have a higher than average density of galaxies. Lapparent, Geller & Huchra (1986) show a redshift plot for a region of the sky that includes the Coma cluster. Their data is from the Center for Astrophysics redshift survey and their plot clearly shows several voids. They suggest that the galaxies are distributed on the surfaces of shells. However, this distribution could also arise from the effects of curvature-redshift in clouds of gas.

6.9.2 Cluster of Galaxies: conclusion

In conclusion, it has been shown that for the Coma cluster, and by implication for other clusters, curvature-cosmology can explain the velocity dispersion of the galaxies without requiring the occurrence of dark matter. The peculiar velocities are due to the effects of curvature-redshift in the intra-cluster gas. The predicted redshift dispersions are one or more magnitudes larger than the redshift due to actual velocities. There is very strong support for this explanation from the observations of redshifts in the Virgo cluster. There is other evidence for curvature-redshift in that redshifts of objects seen through the cluster are larger than expected. A consequence is that much of the structure seen in the velocity distribution of galaxies may be due to curvature-redshift. The supposed voids, walls, and other structures could be the result of large gas clouds. Thus, analysis of velocity dispersions in clusters of galaxies does not provide any support for the existence of dark matter. Finally, the energy loss from curvature-redshift could help explain some of the X-ray luminosity observations in clusters of galaxies.

Observations of clusters of galaxies provide strong support for the hypothesis of curvature-redshift. Since curvature-redshift is a major component of curvature-cosmology, the observations show strong support for curvature-cosmology.

6.10 The Hubble constant

Recently Riess et al. (2005) have measure accurate Cepheid distances to two galaxies containing nearby supernovae. Together with two earlier measurements, they derive an absolute magnitude of -19.17±0.07 mag for the reference supernovae. The average absolute magnitude of the supernovae with $z > 0.5$ in Figure 10 is -19.07±0.09 mag where the uncertainty includes the error in the reference magnitude added in quadrature. Hence, from equation (81) the reduced Hubble constant is given by

$$19.11 + 5\log(h) = 19.07 + 5\log(0.7) \tag{122}$$

that results in a Hubble constant of $H=69.0\pm0.1$ km.s^{-1}.Mpc^{-1}.

As already mentioned analysis of the Coma cluster provides another estimate of the Hubble constant. Allowing for the average velocity of the galaxies due to curvature-redshift in the intra-cluster gas the average velocity (section 6.9) is 6926-1206=5720 kms^{-1} and with a distance of 87.1 Mpc we get 5270/87.1=65.7 km.s^{-1}.Mpc^{-1} which is not inconsistent with the previous value.

The background X-ray results (above) gave a density of 1.34 ± 0.02 m^{-3} and therefore by equation (63) the predicted Hubble constant is $H=48.9\pm0.4$ km.s^{-1}.Mpc^{-1} where the uncertainties are statistical. Curvature-cosmology with just one free parameter, the average density of the universe, has provided an estimate that is in reasonable agreement with the observed value of the Hubble constant. The major systematic uncertainties come from the effects of density fluctuations in the curvature-cosmology model and the effects of curvature-redshifts in gas clouds on the observed redshifts of distant objects.

Since redshift is a function of both the distance and the density of the cosmic plasma the question arises as to its capability as a distance measure. In addition to curvature-redshift due to the cosmic plasma, there are extra redshifts due to the halo of our own Galaxy and the possible halo around the observed object. For large distances, say with redshifts greater than about $z=0.1$, the variations in density should average out and redshift should be a good distance measure. For closer distances, a redshift may differ from the expected value due to the statistics of the small number of clouds passed through in the line of sight. The best estimate of the effect of these clouds comes from the observed inter-galactic gas in clusters of galaxies. It is shown in section 6.9 that the extra redshift due to these clouds is of order (in velocity units) of 1000 km.s^{-1}. Thus, any redshift that is less than about $z\sim0.03$ (~10,000 km.s^{-1}) is probably not a good measure of distance.

What curvature-cosmology has done is to relate the Hubble constant to the average density of the universe. In Big-Bang cosmology and similar expansion cosmologies, there is no a priori reason why such a relationship should occur. Thus, the closeness of the agreement provides very strong support for curvature-cosmology.

6.11 Lyman-alpha forest

The Lyman-alpha (Lyα) forest is the large number of absorption lines seen in the spectra of quasars. Most of the lines are due to absorption by clouds of neutral hydrogen in the line of sight to the quasar. Some of the lines are due to other elements or due to Lyman-beta absorption. Because of the redshift between the absorbing cloud and us, the lines are spread out over a range of wavelengths. Usually the analysis is confined to lines between the Lyα (at a wavelength of 121.6 nm) and Lyβ (at 102.5 nm). Thus, each quasar provides a relatively narrow spectrum of Lyα lines at a redshift just less than that for the quasar. Since the advent of spacecraft telescopes, in particular the Hubble Space Telescope, which can observe the ultraviolet lines, by using many quasars the complete redshift range up to the most distant quasar has been covered. The large redshift range makes the Lyman-alpha spectra potentially a powerful cosmological tool.

The obvious cosmological observation is the density of lines as a function of redshift but as discussed by Rauch (1998) in an excellent review, there are many important observational problems. The first, which has now been overcome, is that the spectra must have sufficient resolution to resolve every line. The second is that most lines are very weak and the number of resolved lines can depend greatly on the signal to noise ratio. This is accentuated because the steep spectrum for the density of lines as a function of their strength means that a small decrease in the acceptance level can drastically increase the number of observed lines. The third problem is that each quasar only provides a narrow range of redshifts and there are considerable difficulties in getting uniform calibrations. In addition to these problems, it will be shown that curvature-redshift can have a profound effect on the interpretation of the line widths and column densities.

Since in curvature-cosmology the distribution of clouds is independent of time or distance the expected density of lines of the same strength as a function of redshift is

$$\frac{dn}{dz} = \frac{acn_0}{H(1+z)}, \qquad (123)$$

where n_0 is the volume density and a is the average area of a cloud. Most observers have fitted a power law with the form $(1+z)^\gamma$

Lyman-alpha forest

with a wide range of results. They vary from $\gamma=1.89$ to $\gamma=5.5$ (Rauch 1998). All of which are inconsistent with curvature-cosmology prediction of $\gamma=-1$. However, Rauch (1998) points out that the majority of lines are always close to the threshold and problems with line blending and data quality can make large differences to the normalisation needed for each quasar. Since each quasar only contributes to a narrow range of redshifts these estimates of γ are very dependent on the normalisation problems between different spectra.

In curvature-cosmology, there is the additional effect that much of the line broadening may be due to curvature-redshift. Curvature-redshift will be operating within the clouds so that the observed line width will be a combination of the usual Voigt profile and the change in the effective central frequency as the photons pass through the cloud. If the cloud has a density $\rho(x)$ at the point x, measured along the photon trajectory then the change in frequency from the entering frequency due to curvature-redshift is

$$\frac{\Delta v}{v} = \frac{1}{c}\int\sqrt{8\pi G\rho(x)}dx. \qquad (124)$$

In units of $N(x)=\rho(x)/m_H$ this is (with N in m^{-3} and dx in kpc)

$$\frac{\Delta v}{v} = -\frac{\Delta\lambda}{\lambda} = \int 1.724\times10^{-7}\sqrt{N(x)}dx. \qquad (125)$$

Then the final profile will be the combination of the natural line width, the Doppler width due to temperature, any width due to bulk motions and the redshift-curvature width. Now assuming pure hydrogen, the hydrogen column density is given by

$$N_H = \int N(x)dx. \qquad (126)$$

Although it is unlikely that the line of sight goes through the centre of the cloud, it is reasonable to expect a roughly symmetric distribution of gas with a shape similar to a Gaussian. We can define an effective density width by

$$x_w^2 = \int (x-\bar{x})^2 N(x)dx / \int N(x)dx. \qquad (127)$$

Also define $n_0=N_H/x_w$ and an effective velocity width Δv, where $\Delta v = 51.68\eta x_w\sqrt{n_0}$ and where η is a small numeric constant that depends on the exact shape of the density distribution. Eliminating the central density, we get (with x_w in kpc)

$$\Delta v^2 = 8.656 \times 10^{-17} \eta^2 N_H x_w \qquad (128)$$

For values $N_H = 10^{19}$ m^{-2} and $x_w = 1$ kpc we get $\Delta v = 29$ km.s^{-1} (with $\eta = 1$). Since there is a wide variation in column densities and the effective widths are poorly known it is clear that curvature-redshift could completely dominate many of the Lyman-alpha line widths and the others would require a convolution of the Doppler profile with the redshift-curvature density effect. What is also apparent is that the very broad absorption lines may be due to curvature-redshift acting in very dense clouds. Although there is uncertainty about the observed relationship between the line width and the column density, we note that for a fixed effective density width, equation (128) predicts a square relationship that may be compared with the exponent of 2.1±0.3 found by Pettini et al. (1990). Clearly, there needs to be a complete re-evaluation of profile shapes, column-densities, and cloud statistics that allows for the effects of curvature-redshift. We must await this analysis to see whether the Lyman-alpha forest can provide a critical test of curvature-cosmology.

7 Evolution

7.1 Evolution in curvature-cosmology

As mentioned previously, curvature-cosmology is static and homogeneous and therefore the statistical properties of the universe must be the same at every point and for every time. Thus, conclusive proof of evolution would invalidate curvature-cosmology. On the other hand, in Big-Bang cosmology evolution is often evoked to explain the apparent changes in the density or magnitude of objects as a function of redshift. In particular, it is claimed that galaxies, quasars, and radio sources show evidence for strong evolution. In this chapter, observations of these claims are disputed and it is shown that curvature-cosmology can explain all of the observations without evolution. To do this we will examine four major topics where evolution is claimed to be important. First, we examine recent data on the properties of galaxies and show that in curvature-cosmology, their absolute magnitudes have a well-defined density distribution and the claimed evolution arises from strong selection effects.

Next, the magnitude distribution of quasars is examined with qualitatively similar results to those for galaxies. Not only do quasars have single peaked density distribution but also any apparent evolution is due to strong selection effects.

The third topic concerns the Butcher-Oemler effect. This is an apparent increase in the fraction of blue galaxies in clusters of galaxies as a function of increasing redshift. The main difficulty is that it is basically a qualitative effect and it is very dependent on the wavelengths that are observed. Although there is not a consensus, many authors believe that the Butcher-Oemler effect is either spurious or at least not proven.

The fourth topic is the number distribution of radio sources as a function of frequency and flux density. It is shown that in curvature-cosmology the observed distribution can be explained with a relatively simple intrinsic distribution of sources that is the same everywhere. Again, there is no evidence of evolution.

7.2 Galaxy distributions

7.2.1 Observations and analysis

In Big-Bang cosmology, the investigation into evolution of galaxies has a long and chequered history. The concept of evolution has progressed from early notions that all galaxies were formed shortly after the Big-Bang to more modern ideas that include the reformation of galaxies in galactic mergers. The basic problem has been to distinguish the evolution of individual galaxies from the average evolution of all galaxies as a function of redshift. In curvature-cosmology, the universe obeys the perfect cosmological principle and any unambiguous evidence that showed that the average characteristics of galaxies varied with distance or time would seriously undermine the theory. What will be shown is that strong selection effects coupled with a dependence of galaxy characteristics on absolute magnitude make the determination of evolution very difficult. Nevertheless, what evidence there is does not show any evolution of the average characteristics of galaxies.

Recently large telescopes with wide fields and the use of many filters have enabled a new type of galactic survey. The light-collecting capability of the large telescopes enables deep surveys to apparent magnitudes of 24 mag or better and the wide field provides a fast survey over large areas. The major innovation is the use of many filters whose response can be used to classify the objects with great accuracy. Thus, galaxies can be separated from quasars without needing morphological analysis. This photometric method of analysis works because photometric templates are available for a wide range of types of galaxies and other types of objects. In addition, accurate redshifts are obtained from fitting the templates without the tedious procedure of measuring the spectrum of each object.

A typical example of this photometric method is the COMBO-17 survey ('Classifying Objects by Medium-Band Observations in 17 filters) provided by Wolf et al. (2004). The goal of this survey was to provide a sample of ~50,000 galaxies and ~1000 quasars with rather precise photometric redshifts based on 17 colours. In practice, such a filter set provides a redshift accuracy of σ_z galaxy ~0.03 and σ_z quasar ~0.1. The central wavelength of the 17 filters varied from 364 nm to 914 nm and consisted of 5 broadband filters (U, B, V, R, and I) and 12 narrower-band filters. The R band filter had a central wavelength of 652 nm and a width (FWHM) of 163 nm and the selection limits were based on the apparent R band magnitude. The detailed analysis uses all the filter outputs to provided redshifts and estimates of the rest frame R band magnitude. Thus, the explicit use of K-corrections was avoided. In addition, each galaxy could be classified by its Spectral Energy Distribution (SED). Wolf et al. (2003) and Bell et al. (2004) claim that there is clear evidence for evolution of the galaxy luminosity function that depends on SED type. Here we will re-examine the COMBO-17 galaxies using curvature-cosmology to compute the absolute magnitudes and show that the results are consistent with a steady state universe.

The basic data for each object in the COMBO-17 survey has been provided by Wolf et al. (2004) on their web site. The data set provides about 63,501 objects observed in an area of 31'.5×30' in the Chandra Deep Field South (CDFS). Following the precepts of Wolf et al. (2003) the redshift range is restricted to 0.2<z<1.1 and the apparent magnitude cut-off is set at an r band apparent magnitude of 23.5 mag. From the published catalogue 6,472 objects were selected that were classified as certain galaxies, with phot_flag < 8, and with 16<Ap_Rmag<24. For this analysis, we assume that H=70 km.s^{-1}.Mpc^{-1} for which the radius of the universe is R=6.06 Gpc (equation (80)).

Since the galaxies were selected by r band apparent magnitude it is necessary to have the R band absolute magnitudes in order to determine for each galaxy the spatial volume in which it could be detected. Ideally, we need the K-correction for each galaxy for all redshifts. Since this data is not available, it was necessary to assume a common R band K-correction for all galaxies. A further approximation was needed in that only the r band apparent magnitude is provided. However, the data set does contain the absolute magnitudes for the SDSS (Sloan Digital Sky Survey) r band that is close to the COMBO-17 selection band. The average difference between these two magnitudes less the Big-Bang distance measure that had been used was obtained as a function of redshift with a step of 0.1 in the redshift parameter z. This was smoothed to reduce observational noise and it was then be used to provide K-corrections for the R band.

The method used to obtain the galaxy density distribution is first to compute for each galaxy its absolute magnitude and then to add $1/V$ to the appropriate magnitude bin where V is the total volume in which the galaxy could be located and still be included in the analysis (Schmidt 1968). One can view the process as measuring density for each galaxy and then adding these densities to obtain the overall density. The uncertainty in the density has one contribution from the variance of $1/V$ and another from the Poisson statistics of the number of galaxies occurring in each bin. Since these two effects are uncorrelated, it can be shown that the estimated standard deviation of the density where there are n galaxies in the bin is given by

$$\sigma_{density} = \sqrt{\frac{1}{n}\sum_{i=1}^{n}\left(\frac{1}{V_i}\right)^2}. \tag{129}$$

The COMBO-17 survey data provides absolute magnitudes for the Johnson U, B, and V bands for the CDFS field. The curvature-cosmology absolute magnitudes were derived by replacing the Big-Bang distance modulus by the curvature-cosmology distance modulus. Figure 17 shows the density distribution as a function of absolute magnitude for the three colours U, B, and V. To reiterate the accessible volume for each galaxy was determined from its r band magnitude, however the bins into which the volume estimates were placed were based on the given absolute magnitudes for each of the three bands. Because of the large number of galactic types, each with a different spectrum,

Figure 17 can provide only a broad average of the galactic densities. Nevertheless, it clearly shows a peaked spectrum that is relatively narrow. A Gaussian curve has been fitted to these curves with the functional form

Galaxy distributions

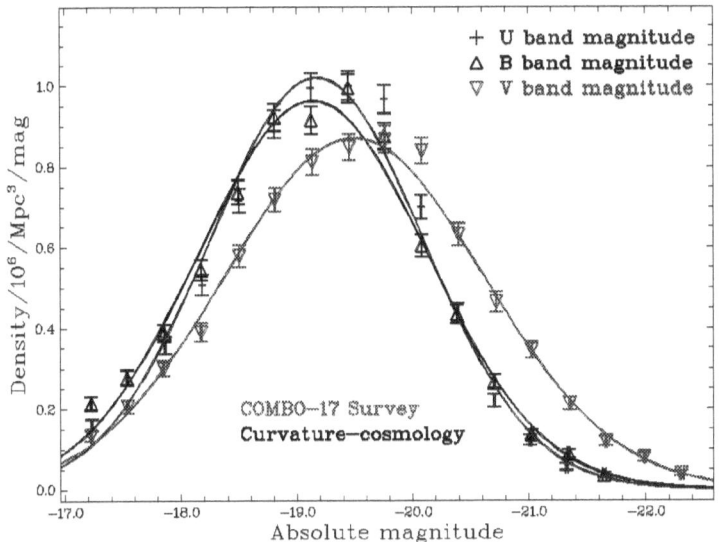

Figure 17: Density of COMBO-17 survey galaxies as a function of absolute magnitude for U, B, and V colour bands.

$$\rho(M) = \frac{\rho_0}{w}\sqrt{\frac{\ln(2)}{\pi}} \exp\left(-\ln(2)\left(\frac{M - M^*}{w}\right)^2\right), \qquad (130)$$

where $\rho(M)$ is the total space density, M^* is the position of the peak and w is the HWHM (half-width-half-maximum) of the distribution. The justification for this form is an appeal to the central limit theorem and that it has a good fit. This expression implies that the density distribution as a function of luminosity is a log-normal distribution. The fitted parameters are given in Table 16.

Table 16: Fitted parameters for galaxies

	Density /10^6/Mpc^3	Peak /mag	HWHM /mag
U	(2.36±0.03)	-19.17±0.01	1.09±0.01
B	(2.39±0.03)	-19.14±0.01	1.16±0.01
V	(2.45±0.03)	-19.49±0.02	1.32±0.02

7.2.2 Galaxy evolution

The basic problem of selection of galaxies to investigate evolution is illustrated in Figure 18. It is essential to confine any analysis of evolution to those galaxies that are visible out to the maximum redshift. Otherwise, the varying cut-off of the tail of the distribution will seriously compromise the evolution analysis. In this case, the cut off lies near the maximum magnitude. Thus, about half the fainter galaxies are excluded from the evolutionary analysis.

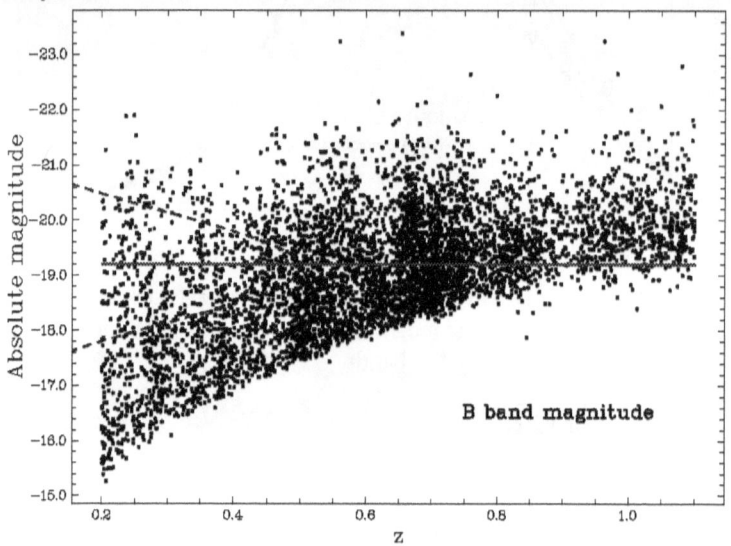

Figure 18: Scatter plot of B band absolute magnitudes verses redshift for COMBO-17 galaxies. All galaxies above the horizontal (solid) line are visible at all redshifts up to z=1.1. Dashed line shows B band density distribution shown above.

In addition, we note that the distribution is not uniform in redshift. There is evidence of several clusters of galaxies and a significant void near $z=0.9$. Thus, any investigation into density evolution is seriously compromised by the uneven distribution. Since the COMBO-17 table does not contain details of galactic type (SED) we are reduced to examining the average absolute magnitude of all galaxies that are bright enough to be visible at $z=1.1$. The cut-off magnitudes for each band are $U_{cut}=-19.6$, $B_{cut}=-19.2$ and $V_{cut}=-19.2$ mag.

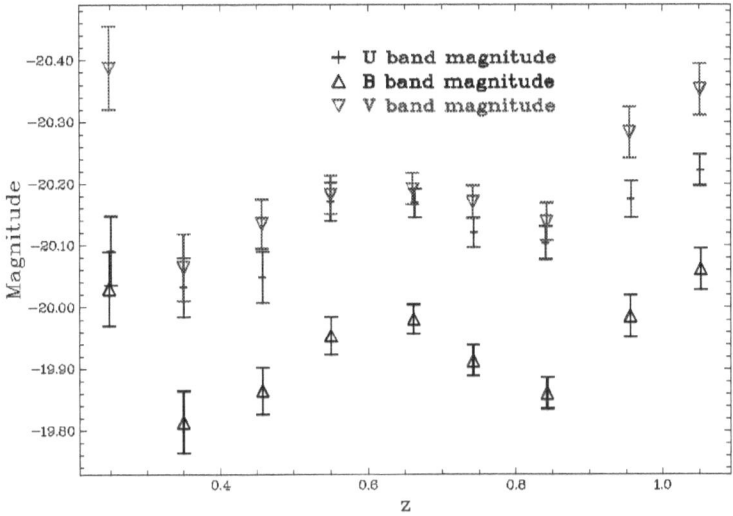

Figure 19: Average absolute magnitudes of galaxies
visible up to z=1.1 verses redshift.

The average absolute magnitudes as a function of redshift are shown in Figure 19. Clearly all the variation could be due to selection effects and there is no systematic evidence for evolution. Ideally, this analysis should be repeated with selection based on galactic type so that the same object can be examined at all redshifts. Nevertheless, it is apparent that there is no evidence of evolution, and the data is fully consistent with curvature-cosmology.

7.3 Quasar distribution

Because they are very bright, quasars can be seen for large distances in space. This makes them very useful for investigating cosmological models. In this regard, it is important to know the distribution of their absolute magnitudes. It will be shown that the quasar data is consistent with curvature-cosmology. There is also the new and important finding that the density distribution of absolute magnitudes is not open-ended but shows a definite peak at about -23 mag.

Recent extensive work by a large number of astronomers has resulted in the production of two large quasar catalogues. There is the Sloan Digital Sky Survey (SDSS) Quasar Catalogue

(Schneider et al. 2005) that has 42,420 objects, of which most are quasars, in a redshift range up to 5.41. Each object has been observed in five bands (ugriz) and has an accurate position and redshift. The second survey comes from the 2dF QSO Redshift Survey of 44576 colour-selected objects (Croom et al. 2004) which were observed in three bands ub_jr (where there is the possibility of confusion with the SDSS bands we have designated their bands u'b'r'). There were 23338 identified quasars that make up the 2QZ survey. In addition, they report on quasars selected from the 6dF QSO Redshift Survey that had 1564 brighter objects that produced 322 quasars to give the 6QZ Survey. The essential parameters for the 2QZ survey, the 6QZ survey, and the SDSS survey are given in Table 17. The selection criteria used in the following analysis were that the Galactic latitude is greater than 5°, redshift selection as shown in Table 17. To be consistent with Croom et al. 2004 and Schneider et al. 2005 it is assumed that $H=70$ km.s^{-1}.Mpc^{-1} for which the radius of the universe is $R=6.06$ Gpc.

The surveys had different selection criteria. All start with an input catalogue of possible candidates. The initial catalogue for the 2QZ+6QZ survey was derived from machine measurements (Automated Plate Measuring Machine (APM) at the Institute for Astronomy, Cambridge, Kibblewhite et al. 1984) of UKST photographic plates of objects with 16<b'<20.85 mag. Candidates were selected for 2QZ if they had 18.25<b'<20.85 mag and satisfied one of the following colour criteria: u'-b'≤-0.36; u'-b'≤0.12-0.8(b'-r'); b'-r'<0.05. For the 6QZ sample with b'<18.25 mag the u'-b' limit was tightened to u'-b'≤-0.50. Spectra were obtained for nearly all of the chosen objects and quasars were identified by matching templates and requiring broad emission lines (>1000 km.s^{-1}). Croom et al. (2004) consider a wide range of factors that could affect the completeness of the survey. These range from morphological completeness where true point sources are misclassified by the APM analysis software as being extended through photometric completeness to spectroscopic completeness. In order to achieve a photometric completeness greater than 85% they restrict the redshift range to 0.4<z<2.1 and use an absolute magnitude limit of M_i>-22.5 mag derived from a Big-Bang cosmology.

The catalogue used for the SDSS survey is the SDSS Third Data Release of quasars that have a luminosity larger than M_i=-22.0 mag (for a Big-Bang cosmology; H=70 km.s^{-1}.Mpc^{-1} and Ω_M=0.3) and whose spectra contain at least one broad (velocity FWHM larger than ≈1000 km.s^{-1}) emission line. There are the additional apparent magnitude limits of 20.2>i>15.0. Schneider at al. (2003) state that the design goals of 65% efficiency and 90% completeness should be achieved. However, they also state that the SDSS quasar survey does not represent a uniform and homogeneous sample and significant effort will be required to quantify this to allow it to be used for statistical purposes.

Table 17: Parameters for the quasar surveys.

Parameter	Unit	2QZ	6QZ	SDSS
Effective area	deg^2	595.9	330.0	4188
Magnitude range	mag	18.25<b'<20.85	16<b'<18.25	M_i<-22 20.2>i>15.0
Number selected		16561	322	42062
Redshift range		0.2<z<2.1	0.2<z<2.1	0.1<z<5.0
Surface density	deg^{-2}	27.8	1.0	10.0

The photometric system for the SDSS survey is based on an updated version of the spectro-photometric ABv system which Fukugita et al. (1996) designate as AB95. One advantage of this system is that the apparent luminosity is a direct integration of the filter response times the luminosity of the object with respect to frequency. Thus, an object that has a constant luminosity per unit frequency step would have the same magnitude in all filter bands.

The reference spectrum was the composite quasar spectrum provided by Vanden Berk et al. (2001) with the spectrum extended to 11000Å using a power law with a slope of -1.58. The filter responses come from the UK Schmidt web site for the u'b'r' bands and from Strauss and Gunn (2001) for the ugriz bands. The latter filter responses already have the atmospheric absorption included. The u'b'r' filter responses were modified to include 1.3 atmospheres of absorption. The calculation of filter responses closely follow the method of Schneider, Gunn & Hoessel (1983)

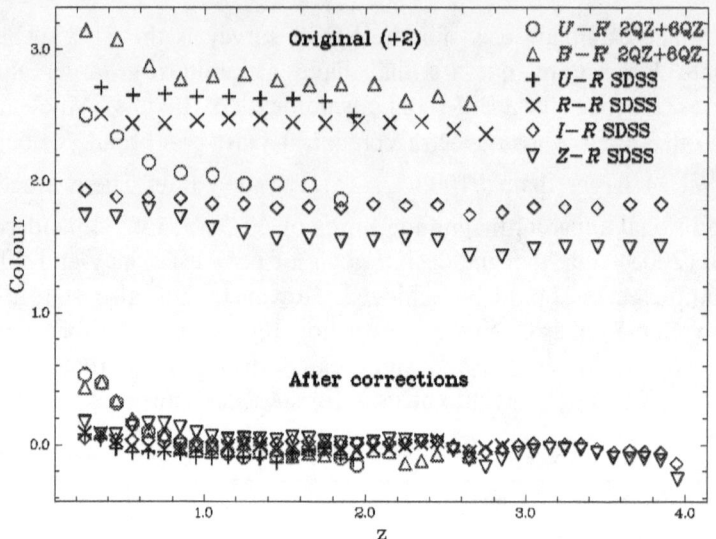

Figure 20: Colour differences for 2QZ+6QZ and SDSS quasars verses redshift. The original differences are at top (displaced by 2 mag).

in which it is assumed that the detectors are photon counters. We can then use the given uncertainties (for the SDSS data) to compute a weighted mean and standard deviation. The method of analysis that was used for both surveys is the use the K-corrections for each quasar to transform the apparent magnitude in each band to an effective r band response. Then the average response and its estimated standard deviation could be computed. Data were used only from bands in which the filter response was fully within the redshifted reference spectrum.

If the K-corrections are ideal then the expected absolute magnitude colour differences should be zero. The original colour differences are shown in the top part of Figure 20 and it is obvious that there are systematic errors in the values. In addition to possible errors in the K-corrections there appear to be errors in the relative gains of the filters. Since it is difficult to disentangle these two forms of error a combined approach has been adopted.

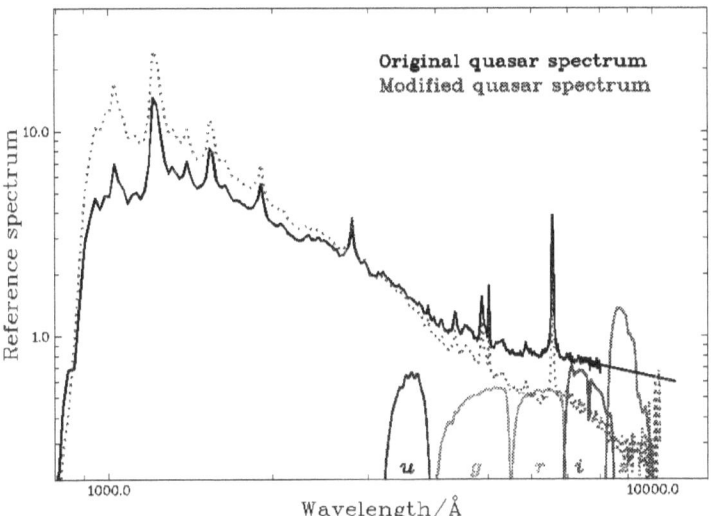

Figure 21: Reference quasar spectrum and filter response curves.

The procedure used was to compare the response for each filter for quasars from the SDSS survey after the K-corrections were applied and to assume that the difference between its response and that for the r band was due to an incorrect reference spectrum. Then a correction was applied to the reference spectrum at the redshifted wavelengths of that filter that was equal to the gain difference multiplied by the appropriate filter response function. This was done for all the SDSS quasars and resulted in a new corrected reference spectrum. The 2QZ+2QZ survey quasars were not used for these corrections because there was data for only three filters and there is uncertainty as to whether a photographic plate should be treated as a photon or energy detector. The next step was to compute a filter gain correction by requiring that the average value of the absolute magnitude from all the quasars should be the same. Since this can only provide relative gains the r' filter was adopted as the reference for 2QZ+6QZ survey and the r filter was adopted as the reference for the SDSS survey. The process was iterated until there was no significant change in the spectrum. The original and corrected reference spectra are shown in Figure 21. The noise at the long wavelength end of the modified spectrum has arisen from the corrections due to the edge of the z filter varying with quasars with different redshifts. Because of the wide filters, there is little change to the emission features and therefore the modified spectrum is mainly useful in computing K-corrections and is not a good estimate of the true spectrum. Figure 21 also shows the response curves (with arbitrary normalisation) for the SDSS filters. The results for the filter gains are shown in Table 18 where the wavelength column shows the geometric mean wavelength, the gain column is the average base level (in mag) for each filter and the rest frame response is the integration of each filter response for the corrected spectrum at zero redshift.

Table 18: Filter characteristics

Filter	Wavelength/Å	Gain/mag	Rest frame response
u'	3601.7	-0.076	-1.497
b'	4468.8	0.705	-0.774
r'	6551.2	0.0	0.086
u	3540.1	0.571	-1.569
g	4715.8	0.386	-0.626
r	6193.8	0.0	0.0
i	7566.1	-0.255	0.518
z	8907.3	-0.302	0.068

When this was done the average estimated standard deviation in the corrected magnitude for the quasars in the 2QZ+6QZ survey changed from 0.140 ± 0.002 mag to 0.134 ± 0.002 mag and for the SDSS survey it changed from 0.055 ± 0.002 mag to 0.052 ± 0.002 mag. It should be emphasised that these corrections are completely independent of any cosmological model. Although the changes in the reference spectrum shown in Figure 21 appear to be large, the major result is small changes in absolute calibrations. The modified reference spectrum was used because it is desirable to remove any systematic effects of calibrations that may confuse evidence of evolution.

An important step in the analysis is to determine as a function of redshift the magnitude of the faintest quasar that would be included within the survey. This function is also used to determine the volume of space in which a quasar with a given absolute magnitude could be observed.

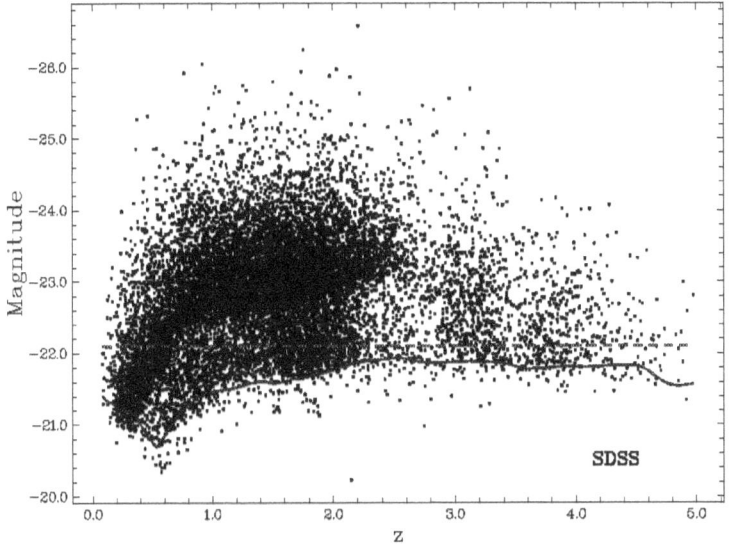

Figure 22: Scatter plot of absolute R magnitudes of SDSS quasars verses redshift.

Figure 22 shows a scatter plot of the absolute (R) magnitudes of the SDSS quasars. The horizontal dashed line shows those quasars that can be observed up to a redshift of five and the lower solid line is the faint magnitude limit. It has been computed from the absolute magnitude and apparent magnitude limits used in the SDSS survey. First, the apparent magnitude limit (m_i) for each quasar was computed for an absolute magnitude of -22.0 mag using the distance measure computed for a Big-Bang cosmology with H=70 km.s^{-1}.Mpc^{-1} and Ω_M=0.3. Then their K-correction for a power-law-frequency-spectrum with a power of -0.5 was applied. The limiting apparent magnitude is equal to m_i if it is brighter than the apparent magnitude limit of 20.2 mag. The change-over occurs in the r band at a redshift of about 0.6. The other variations in the limiting magnitude are due to the effects of the K-correction in going from the i band to the r band. The absolute magnitudes shown in Figure 22 are computed for curvature-cosmology.

The equivalent figure for the 2QZ+6QZ survey data is similar. Finally, there is a need to take into account the fact that for the 2QZ+6QZ survey the quasars come from two surveys with different sky areas. Since the selection of which of these surveys a quasar comes from is based on its apparent magnitude, and in this case, the two surveys have non-overlapping magnitude ranges, they are mutually distinct and it is therefore appropriate to compute the quasar densities for each survey and add them. The analysis method used to obtain quasar densities is essentially identical to that used for the galaxies. The only difference is that each quasar is assumed to have the same spectrum. The process is to compute for each quasar its absolute magnitude and then to use it to calculate the total volume in which the quasar could be located and still be included in the analysis (Schmidt 1968, also see section 7.2). The result from both surveys is shown in Figure 23 with both surveys using curvature-cosmology for estimating the distance measures and the volumes. Both surveys show a well-defined density peak at about -23 mag. This is different from traditional Big-Bang analysis that does not show a peak but a monotonic increase of density for fainter galaxies (Croom et al. 2004). The solid curves are the best fitting Gaussian curves with the functional form given by equation (130).

Table 19: Parameters for quasar density fit

	Density/10^6/Gpc^3	Peak(M*) /mag	HWHM (w) /mag
2QZ+6QZ	3.94±0.04	-22.454±0.007	0.801±0.006
SDSS	1.46±0.01	-22.817±0.004	0.854±0.003

The parameters for the two surveys are given Table 19. The uncertainties shown in Table 19 are the statistical errors from the fitting procedure. The systematic uncertainties are very much larger. This difference in total densities by a factor of 2.7 is also apparent in the surface densities shown in Table 17 where the ratio is 2.9. Presumably, it is due to different efficiencies of finding quasars or may be due to a mistake in the quoted area of either survey. The differences in the peak magnitudes is probably due to calibration errors and since the 2QZ+6QZ survey is based on photographic plates it should have a different reference spectrum and K-corrections from those appropriate for the photon detectors in the SDSS survey.

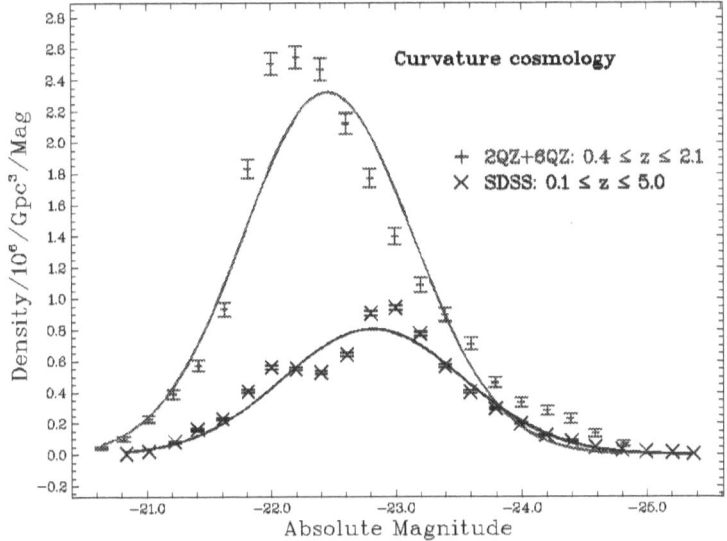

Figure 23: Density distribution of quasars as a function of absolute magnitude.

The conclusion is that quasars have a density distribution that is approximately Gaussian in shape with a peak at an absolute magnitude of -23.0 and a HWHM value of 0.7 mag. Strictly speaking, because of their different spectra; we cannot compare the quasar distributions to the galaxy distributions. However, a rough comparison shows that the typical quasar is about 30 times more luminous than the typical galaxy.

7.3.1 Quasar evolution

The major purpose of this section is to investigate whether there is any evidence of evolution in the quasar data. Examination of Figure 22 shows that a large number of quasars are visible at all redshifts within the range of the survey. The horizontal straight line at -22.1 mag shows the boundary. A similar boundary occurs in the 2QZ+6QZ survey at -21.9 mag. If there is no evolution, the properties of these quasars should not vary with redshift. The obvious property is the spatial distribution but as can be seen in Figure 22 this distribution is far from uniform. There is a significant deficiency in quasar numbers near a redshift of 2.5 and at lower redshifts, there appears to be a separation into brighter and fainter groups. These variations are easily explained by the colour selections that have been used in selecting the quasars. The 2QZ+6QZ survey does not show these density variations but suffers in this context from having a much smaller redshift range.

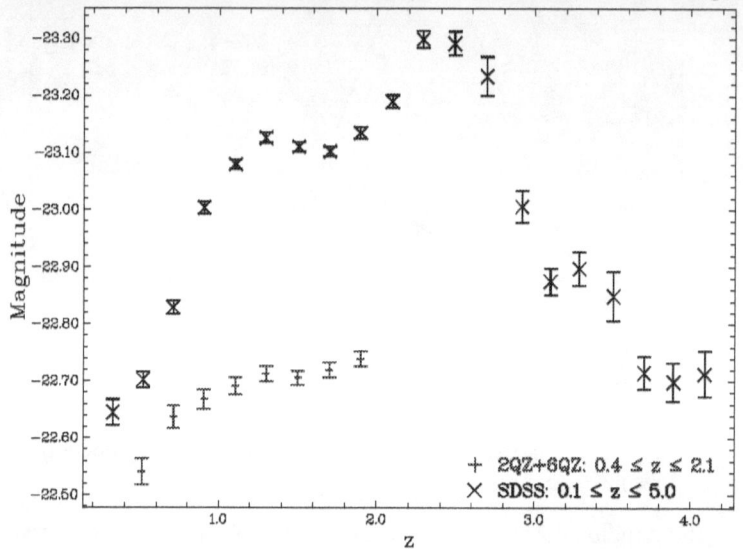

Figure 24: Average absolute magnitude of quasars visible out to z=4 verses redshift.

Another quasar property that should be independent of redshift is the average absolute magnitude. Figure 24 shows the average absolute magnitude as a function of redshift for both surveys. Although there is an obvious variation, the crucial point is that the variation is very small with the total variation being about one mag. The observed variations could easily be due to the quasar selection effects and it is unlikely that they are due to evolution.

The purpose of this exercise was to show that the quasar observations are consistent with curvature-cosmology rather than to do a detailed investigation of quasars. If curvature-cosmology is valid then there are still important problems in increasing the efficiency of quasar selection. What is needed is a complete count of fainter quasars out to a redshift of about three rather than observations to much larger redshifts. However, current quasar surveys are consistent with curvature-cosmology and show no evidence of redshift evolution.

7.4 The Butcher-Oemler effect

If there were evidence of significant change in the universe as a function of redshift, it would be a detrimental to any static cosmology. Probably the most important evidence for this cosmic evolution that is independent of any cosmological model is the Butcher-Oemler (1978) effect. Although they had discussed the effect in earlier papers, the definitive paper is Butcher & Oemler (1984). They observed that the fraction of blue galaxies in galactic clusters appears to increase with redshift. Clusters allow the study of large numbers of galaxies at a common distance and out to large redshifts, which makes them ideal for studies in evolution. The core regions in a cluster are dominated by early-type (elliptical and lenticular) galaxies, which have a tight correlation between their colours and magnitudes. We can calculate R_{30}, the projected cluster-centric radius that contains 30% of the total galaxy population. The blue fraction, f_B, is defined to be the fraction of galaxies within R_{30} which are bluer than the colour-magnitude relationship for that cluster. At first sight, this may appear to be a simple test that could be done with apparent magnitudes. However to compare the ratio for distant clusters with that for nearby ones the colours must be measured in the rest frame of each cluster, hence the need to use K-corrections. The major advantage of the Butcher-Oemler effect is that it is independent of the luminosity-distance relationship that is used. Therefore, to be more precise f_B is the fraction that has an absolute magnitude M_V, whose rest frame (B-V) colour is at least 0.2 magnitudes bluer than expected. A review by Pimbblet (2003) summarises the important observations.

In its original form the Butcher-Oemler effect is dependent on the apparent magnitude cut-off limits. It is essential that selection effects are the same in the rest frame for each cluster. There are further complications in that the percentage of blue galaxies may or may not depend on the richness of the cluster and the effect of contamination from background galaxies. Although Pimbblet (2003) concluded that, there was a definite effect, his Figure 1 shows that this conclusion is open to debate. Since then there have been several attempts to measure an unambiguous effect. Even though they attempted to duplicate the original methodology of Butcher & Oemler, De Propris et al. (2003) found essentially no effect for K-selected galaxies. Andreon et al. (2004) examined three clusters around $z=0.7$ and did not find clear-cut evidence for the effect. To quote one of their conclusions: *Twenty years after the original intuition by Butcher & Oemler, we are still in the process of ascertaining the reality of the Butcher-Oemler effect.*

The Butcher-Oemler effect remains uncertain, and therefore does not provide evidence to refute curvature-cosmology.

7.5 Radio Source Counts

The count of the number of radio sources as a function of their flux density is one of the earliest cosmological tests that arose from the development of radio astronomy after World War II. Indeed, this test played a pivotal role in the rejection of the steady state cosmology of Bondi, Gold, and Hoyle in favour of the Big-Bang evolutionary model. In recent years, the study of radio source counts has declined for several reasons both theoretical and experimental. An important experimental problem is that many radio sources are double or complex in structure. Whether or not they are resolved depends on their angular size and the resolution of the telescope. Since their distance is unknown, the counts are distorted in a way that cannot be readily determined. The main theoretical problem in Big-Bang cosmology is that the counts are of a collection of quite different objects such as galaxies and quasars that can have different type of evolution. Thus, the radio source counts are not very useful in the study of these objects. However, in curvature-cosmology, the luminosity number density must be the same at all places and at all times. Thus, curvature-cosmology demands that radio source counts are consistent with a reasonable luminosity number density distribution that provides a critical test of the cosmological model.

In order to clarify the nature of the radio source count distribution let us start with a simple Euclidean model. Let the observed flux density of a source at an observed frequency of v_0 be $S(v_0)$ in units of $W.m^{-2}.Hz^{-1}$ and its luminosity at the emitted frequency v be $L(v)$ in units of $W.Hz^{-1}$. For simplicity, let us

Radio Source Counts

assume that all the sources have the same luminosity and that they have a volume density of N sources per unit volume. Then using the inverse square law, the observed number of sources with a flux density greater than S is

$$n(>S) = \frac{4\pi N}{3}\left(\frac{L}{4\pi S}\right)^{3/2}.$$

Thus the number density of observed sources is $dn/ds = S^{-5/2}$.

The importance of this result is that it is customary to multiply the observed densities by $S^{5/2}$ so that if the universe had Euclidean geometry the distribution as a function of S would be constant. It has the further advantage in that it greatly reduces the range of numbers involved.

For curvature-cosmology the area at a distance r is $A(r) = 4\pi R^2 \sin^2(\chi)$ where $r = \sin(\chi)$ and $\chi = \ln(1+z)/\sqrt{2}$. Note that the actual light travel distance is $R\chi$. Thus

$$S(v_0)dv_0 = \frac{L(v)dv}{4\pi R^2 \sin^2(\chi)(1+z)}, \qquad (131)$$

where $v = (1+z)v_0$ and the $(1+z)$ divisor allows for the energy loss due to curvature-redshift. Since the ratio of the differentials (the bandwidth factor) contributes a factor that cancels the energy loss, the result is

$$S(v_0) = \frac{L((1+z)v_0)}{4\pi R^2 \sin^2(\chi)}. \qquad (132)$$

It is convenient to replace the distance variable by the redshift parameter z. Then the differential volume is

$$A(r)dr = \frac{4\pi R^3 \sin^2(\chi)}{\sqrt{2}(1+z)}dz$$

If the luminosity number density is $N(L,(1+z)v_0)$ the expected radio-source-count-distribution (allowing for the dS/dL term needed to match the differentials) is

$$n(S,v_0) = \frac{16\pi^2 R^5}{\sqrt{2}} \int_0^{z_m} \frac{\sin^4(\chi)}{1+z} N(L,(1+z)v_0)dz, \qquad (133)$$

where $z_m = 8.22$.

A major problem with the observations is the difficulty in knowing the selection criteria. Typically, all sources greater than a chosen flux density are counted in a defined area. Since the flux density measurements are uncertain and the number of sources is a strong function of the flux density, it is difficult to assess a statistically valid cut-off for the survey. In a static cosmology, the change in the distribution due to the change in emitting frequency as a function of z is an added complication. Thus, an essential test of curvature-cosmology is to show that there is an intrinsic distribution of radio sources that is identical at all redshifts. Unfortunately, it is not feasible to obtain a definitive distribution. What will be done is to show that the observations are consistent with a possible distribution. The aim of this section is to show that there is a distribution $N(L,v_e)$ that provides a reasonable fit to the observations at all frequencies. A simple distribution has been found that provides a good first approximation to the intrinsic distribution. Define the variable x by

$$x = \left(\frac{L}{L^*}\right)\left(\frac{v_0(1+z)}{1\,\text{GHz}}\right)^\beta, \qquad (134)$$

where L^* and β are constants. The second term in equation (134) is the only frequency contribution in this simple model. Then the model for the intrinsic radio-source-distribution is

$$N(L,v) = Ax^{-\alpha}\exp(-\gamma x^2). \qquad (135)$$

where α, β, γ, and A are constants that are found by fitting the model to the data listed in Table 21. In order to provide realistic values a value of $R=10$ Gpc has been adopted. In addition, the integration was started at $z=0.07$ in order to avoid some convergence problems. The results of fitting this distribution to the data are shown in Table 20. The χ^2 goodness of fit was 3828 for 250 DoF. Because of the poor fit, the estimate of statistical uncertainties has been omitted.

A plot of the data (references in Table 21) with the flux densities in Jy, and the results of this model is shown in Figure 25. Each data point has been multiplied by $S^{5/2}$ and each set of points for a given frequency has been multiplied by a factor of 10 relative to the adjacent group.

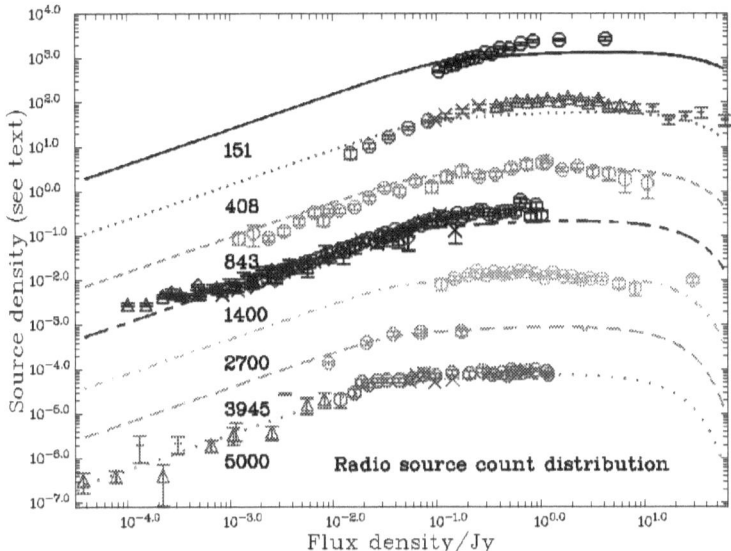

Figure 25: Radio source volume density distribution as a function of flux density. Numbers show frequency (in MHz) of each group. Each point has Euclidean normalisation and each group is displaced from the adjacent group by a factor of 10.

Table 20: Parameters for radio source distribution.

Parameter	Value	Unit
α	1.714	
β	0.334	
γ	0.319	
L^*	6.63×10^3	$W.Hz^{-1}$
A	1.49×10^6	Gpc^{-3}

Clearly, the model satisfies the basic structure of the distributions but there is a poor fit at low frequencies, which is probably due to the limitations of the simple distribution (equation (135)). It should be emphasised that the only free parameters are those shown in Table 20. This analysis shows that a simple distribution of radio source flux densities can be found that is consistent with the observations. In Big-Bang cosmology even with the inclusion of evolution there is no model of radio sources that is as simple as the one described here. Thus, the observations of the number distribution of radio sources as a function of flux density and frequency shows very strong support for curvature-cosmology.

Table 21: Origin of radio source-count data.

Survey name	Telescope	Freq. /MHz	Reference
7C	CLST	191	McGilchrist et al.1990
5c6	Cambridge One-Mile	408	Pearson & Kus 1978
B2	Bologna	408	Colla et al.1973
All Sky		408	Robertson 1972
Molonglo Deep	Molonglo Cross	408	Robertson 1977a,1977b
Molonglo	MOST	843	Subrahmanya 1987
FIRST	VLA	1400	White et al.1997
Virmos	VLA	1400	Bondi et al.2003
Phoenix	ATCA	1400	Hopkins et al.1998
ATESP	ATCA	1400	Prandoni et al.2001
ELIAS	ATCA	1400	Gruppioni et al.1999
Parkes	Parkes	2700	Wall & Peacock 1985
RATAN	RATAN-600	3945	Parijskij et al.1991
	100m (MPIfR)	4850	Maslowski et al. 1984a, 1984b
	VLA	5000	Bennett et al.1983
	VLA	5000	Partridge et al.1986
MG II	NRAO 300´	5000	Langston et al.1990

7.6 Evolution: conclusion

In this chapter, we have examined whether there is definite evidence for evolution in the distribution of galaxies, for quasars, for the Butcher-Oemler effect, and for the distribution of radio sources. It was shown that in curvature-cosmology there were well-defined density distributions for both galaxies and quasars. More importantly, it was shown that the presumed evolution for both galaxies and quasars is explained by strong selection effects. Whether the Butcher-Oemler effect is valid is debatable with evidence for its existence being ambiguous. Indeed none of these observations provided any unambiguous evidence of evolution.

The curvature-cosmology distribution of radio sources is a better characterisation of the observations than has been achieved in Big-Bang cosmology with the inclusion of evolution. Thus, there is strong support for curvature-cosmology with no evidence of evolution that would contravene its basic premise of a static universe.

8 Non-cosmological tests

This chapter considers non-cosmological tests that are direct tests of either curvature-redshift or curvature-pressure. The first section examines whether curvature-redshift can explain the anomalous rotation curves that are seen in spiral galaxies. The second section consists of a discussion of what effects curvature-redshift would have on observations in our own galaxy, the Milky Way. Next, it is shown that curvature-pressure can explain that the detected rate of neutrinos coming from the sun is only about half the rate predicted by the standard solar model. Including curvature-pressure in the model provides a predicted rate that is in excellent agreement with the observations. One of the effects of curvature-redshift is that when radiation traverses plasma the energy lost by the radiation goes into heating the plasma. One possible application of this heating is in the solar corona. It is shown that curvature-redshift is not a significant process in heating the corona. Finally it is shown that curvature-redshift due to inter-planetary dust can explain the anomalous acceleration that has been observed in Pioneer 10 and some other planetary spacecraft.

8.1 Galactic rotation curves

One of the most puzzling questions in astronomy is: why does the observed velocity of rotation in spiral galaxies not go to zero towards the edge of the galaxy. Simple Keplerian mechanics suggests that there should be a rapid rise to a maximum and then a decrease in velocity inversely proportional to radius once nearly all the mass has been passed. Although the details vary between galaxies, the observations typically show a rapid rise and then an essentially constant velocity as a function of radius out to distances where the velocity cannot be measured due to lack of material. The standard explanation is that this is due to the gravitational attraction of a halo of dark matter that extends well beyond the galaxy. We examine whether this rotation curve can be explained by curvature-redshift.

Observations show that our own Galaxy and other spiral galaxies have a gas halo that is larger than the main concentration of stars. It is clear that if the observed redshifts are due to curvature-redshift acting within this halo, the halo must be asymmetric; otherwise, it could not produce the asymmetric 'rotation' curve. Now the observed velocities in the flat part of the curves are typically 100 to 200 km.s^{-1}. The first step is to see if curvature-redshift provides the right magnitude for the velocity. For a gas with an average density of N (effective mass hydrogen atoms per cubic metre) the predicted redshift (in velocity units) is $5.17 \times 10^{-2} d\sqrt{N}$ km.s^{-1} where d is the distance in kpc. Thus for realistic values of d=10 kpc and N=1.0×10^5 m^{-3} the velocity is 163 km.s^{-1}. Thus, the magnitude is feasible.

Although there could be a natural asymmetry in a particular galaxy, the fact that the flattened rotation curve is seen for most spiral galaxies suggests that there is a common cause for the asymmetry. One possibility is that the asymmetry could arise from ram pressure. Since most galaxies are moving relative to the cosmic medium, it is expected that there will be an enhanced density towards the leading point of the galaxy. This asymmetric density could produce an apparent velocity gradient across the galaxy that could explain the apparent rotation curve. Naturally, there would be range of orientations and the apparent velocity gradient must be added to any intrinsic rotation curve to produce a wide diversity of results. Thus, curvature-redshift could explain the galactic rotation curves if there is an asymmetric distribution of material in the galactic halo.

8.2 Redshifts in our Galaxy

In our Galaxy, the Milky Way, there is an interesting prediction. The density of the inter-stellar ionized gas is high enough to inhibit curvature-redshift for radio frequencies. From equation (25) it was shown that for wavelengths longer than about $20.6 N^{-1/2}$ m the effect of refractive index in fully ionised plasma will inhibit curvature-redshift. The refractive index of neutral hydrogen is too low to inhibit curvature-redshift. However, any fully ionised plasma with $N_e > 10^4$ will inhibit curvature-redshift for the 21 cm hydrogen line. Since the local interstellar medium has an electron density of about 10^5 m^{-3} (Redfield 2006), curvature-redshift will be inhibited for the 21 cm hydrogen line near the sun. Thus for sight lines close to the Galactic plane we can assume a similar density and thus a similar inhibition with the result that the observed radio redshifts can be correctly interpreted as velocities. Thus, there is little change needed to the current picture of Galactic structure and rotation derived from 21 cm redshifts. However, there may be some curvature-redshift present in sight lines away from the plane and especially in the Galactic halo.

Since optical redshifts have the full effects of curvature-redshift, it should be possible to find objects with discrepant redshifts where the optical redshift is greater than the radio redshift. The difficulty is that the two types of radiation are produced in radically different environments: the optical in compact high temperature objects, such as stars, and the radio in very low-density cold clouds. In addition, there is the complication that optical extinction due to dust limits the optical range to about one kpc.

Curvature-redshift may help to explain an old stellar mystery. There is a long history provided by Arp (1992) of observations of anomalous redshifts in bright hot stars, which is called the K-term or K-effect. Allen (1976) states that B0 stars typically show an excess redshift of 5.1 km.s^{-1}, A0 have 1.4 km.s^{-1} and F0 have 0.3 km.s^{-1}. This can be explained if these stars have a large corona that produces a curvature-redshift. It is probably no coincidence that such stars have large stellar winds and mass outflows. For these cases, curvature-redshift is computed with equation (11). In order to see if it is feasible let us consider a simple model for the outflow in which the material has a constant velocity v_0, and conservation of matter (Gauss's Law) then requires that the density has inverse square law dependence. Although this is incorrect at small stellar radii, it is a reasonable approximation further from the star. Then if ρ_1 is the density at some inner radius r_1, then integration of equation (11) out to a radius r_2, the expected redshift in velocity units is

$$v = \sqrt{\frac{2G\dot{M}}{v_o}} \log\left(\frac{r_2}{r_1}\right), \tag{136}$$

Where $\dot{M}$ is the observed stellar mass-loss-rate. Then with $\dot{M}$ in solar masses per year, with v and v_0 in km.s^{-1}, the redshift is

$$v = 91.7\sqrt{\frac{\dot{M}}{v_o}} \log\left(\frac{r_2}{r_1}\right) \text{ km.s}^{-1}. \tag{137}$$

With $\dot{M} = 10^{-5} M_\odot$ yr^{-1} (Cassinelli 1979), v_0= 1 km.s^{-1} and $r_2/r_1=10^3$ the predicted redshift (in velocity units) is 2 km.s^{-1} which is in reasonable agreement with the observed K-effects mentioned above.

8.3 Solar neutrino production

Since the Homestake mine neutrino detector started operation in the late 1960's, its observations have shown a deficiency in the observed intensity of solar neutrinos compared to accurate theoretical calculations. This has led to an enormous activity in the development and testing of solar models. Although there is evidence that the discrepancy may be explained by neutrino oscillations where neutrinos with a non-zero rest mass can oscillate between various flavours, it is worthwhile seeing if the inclusion of curvature-pressure can alter the predictions of the solar neutrino production rate. The solar model used here is based on that described by John N. Bahcall (Bahcall 1989). For a local context, curvature-pressure is given by equation (49). What we have done is to use the tables (for solution BS05) generously provided by Bahcall in his web site and used them to calculate curvature-pressure. We then assumed that the thermodynamic pressure was reduced by the value of the curvature-pressure and then we used the thermodynamic pressure as an index into the same tables to get the temperature. This largely avoids all the complications of equations of state and changing compositions. Naturally, this will only work if the corrections, as they are here, are small. Then this temperature was used as an index into the neutrino production table to get the production rate for each of the eight listed reactions. As a calibration and a check, the same program was used to compute the rates with no curvature-pressure.

In this test, the maximum discrepancy from the expected rates was 1.3%.

At a radius of 0.1 solar radii, the reduction in thermodynamic pressure was 12.5% and the reduction in temperature was 4.1%. The computed rates with curvature-pressure included in the solar model are shown in Table 22. (Bahcall 1989, Table 6.6) where the standard rates are from Bahcall, Pinsonneault & Basu (2001). The solar neutrino unit (SNU) is a product of the production rate times the absorption cross section and has the units of events per target atom per second and one SNU is defined to be 10^{-36} s^{-1}. For example for each ^{71}Ga target atom in the detector the expected event rate due to solar neutrinos for the pp reaction would be 57.7×10^{-36} s^{-1}. The last row shows the expected event rates for ^{37}Cl and ^{71}Ga target atoms where the uncertainties are proportional to those provided by Bahcall, Pinsonneault & Basu (2001). Another type of detector uses Cherenkov light from the recoiling electron that is scattered by the neutrino. Because this electron requires high-energy neutrinos to give it enough energy to produce the Cherenkov light this type of experiment is essentially sensitive only to the ^{8}B neutrinos.

Table 22: Computed production rates for solar neutrinos for the standard model including curvature-pressure.

Reaction	Relative rate	Rate /cm^2·s^{-1}	Rate/SNU for ^{37}Cl	Rate/SNU for ^{71}Ga
pp	0.829	4.93×10^{10}	0.0	57.8
pep	0.767	1.07×10^8	0.17	2.15
^{7}Be	0.537	2.56×10^9	0.64	18.4
^{8}B	0.288	1.45×10^6	1.67	3.48
^{13}N	0.503	2.76×10^8	0.045	1.71
^{15}O	0.349	1.68×10^8	0.115	1.91
^{17}F	0.318	1.79×10^8	0.0	0.03
hep	0.905	8.42×10^3	0.036	0.09
Totals			2.66±0.42	85.6±5.4

McDonald (2004) provides a list of recent observational results and they are compared with the predictions in Table 23. The columns show the name of the experiment, the type of detector, the unit, the predicted rate (with curvature-pressure), the observed rate, and the χ^2 of the difference from the predicted value.

The statistical and systematic uncertainties have been added in quadrature to get the observed uncertainty. The result in the last row from SNO is from the charged current reaction (v_e+d→p+p+e) that is the expected rate if there are no neutrino oscillations.

The agreement is excellent. However, there may be some biases that could be either theoretical or experimental in origin. The crucial test requires computation with a solar model that includes curvature-pressure so that the more subtle effects are properly handled. The benefit of this agreement is that it gives very strong support for curvature-pressure in a non-cosmological context.

Table 23: Comparison of predicted and observed solar neutrino production rates.

Experiment	Type	Unit	Predicted (with pc)	Observed	χ^2
Homestake	^{37}Cl	SNU	2.66±0.42	2.56±0.23	0.04
GALLEX+ GNO	^{71}Ga	SNU	85.6±5.4	70.8±5.9	3.42
SAGE	^{71}Ga	SNU	85.6±5.4	70.9±6.4	3.08
Kamiokande	electron recoil	$10^6 cm^{-2}.s^{-1}$	1.45±0.26	2.8±0.38	8.60
Super-Kamiokande	electron recoil	$10^6 cm^{-2}.s^{-1}$	1.45±0.26	2.35±0.08	10.95
SNO (v_e+d)	electron recoil	$10^6 cm^{-2}.s^{-1}$	1.45±0.26	1.76±0.10	0.25

8.4 Heating of the solar corona

For over fifty years, astrophysicists have been puzzled by what mechanism is heating the solar corona. Since the corona has a temperature of about 2×10^6 K and lies above the chromo sphere that has a temperature of about 6000K, the problem is where the energy comes from to give the corona this high temperature. Let us consider whether curvature-redshift due to the gas in the corona can heat the corona via the energy loss from the solar radiation.

Aschwanden (2004) quotes the number distribution of electrons in the corona to be

$$N_e = 2.99 \times 10^{14} h^{-16} + 1.55 \times 10^{14} h^{-6} + 3.6 \times 10^{12} h^{-1.5} \text{ m}^{-3}, \quad (138)$$

where h is the distance from the solar surface in units of solar radii. If we assume spherical symmetry then all the radiation leaving the sun must pass through a shell centred on the sun and we can use equation (11) and equation (138) to compute the fractional energy loss in that shell. To the accuracy required, we can also assume that the hydrogen number density is the same as the electron density and then the integration of equation (138) from the solar surface to 4 solar radii above the surface gives a total fractional energy loss of 1.32×10^{-11}. Thus with a solar power output of 3.83×10^{26} W the total energy loss to the solar corona by curvature-redshift is 5.1×10^{15} W which is equivalent to 8.3×10^{-4} W.m^{-2} at the surface of the sun. This may be compared with the energy losses from the corona to conduction, solar wind and radiation. The total loss rates are quoted by Aschwanden (2004) to be 8×10^2 W.m^{-2} for coronal holes, 3×10^3 W.m^{-2} for the quiet corona and 10^4 W.m^{-2} for an active corona. Since these are about seven magnitudes larger than the predicted loss, curvature-redshift is not important in the inner corona. Although it is not pursued here, there is a similar problem in that the Milky Way has a corona with a high temperature. It is intriguing to speculate that curvature-redshift may explain the high temperature of the galactic halo.

8.5 Pioneer 10 acceleration

Precise tracking of the Pioneer 10/11, Galileo and Ulysses spacecraft (Anderson et al. 1998, 2005) have shown an anomalous constant acceleration for Pioneer 10 with a magnitude $(8.74 \pm 1.55) \times 10^{-10}$ m.s^{-2} directed towards the sun. The major method for monitoring Pioneer 10 is to measure the frequency shift of the signal returned by an active phase-locked transponder. These frequency measurements are then processed using celestial mechanics in order to get the spacecraft trajectory. The simplicity of this acceleration and its magnitude suggests that Pioneer 10 could be a suitable candidate for investigating the effects of curvature-redshift. There is a major problem in that the direction of the acceleration corresponds to a blue shift whereas curvature-redshift predicts a redshift. Nevertheless, we will proceed, guided by the counter-intuitive observation that a drag force on a satellite actually causes it to speed up. This is because the decrease in total energy makes the satellite change orbit with a redistribution of kinetic and potential energy.

Let the observed frequency of Pioneer 10 at a distance r, be denoted by v_{obs}, then since the effect of curvature-redshift is proportional to distance and if the actual velocity is v then $v_{obs}=v+\alpha$ where using equation (10), α is given by

$$\alpha = 2\int_0^r \sqrt{8\pi G\rho(r)}\,dr, \quad (139)$$

where the factor of 2 allows for the two-way trip. For the moment, assume that for the region of interest, beyond 20 AU, the variation in the density of the dust and gas with distance is negligible and therefore the curvature-redshift is proportional to r. Then the effective acceleration is

$$\dot{\alpha} = \sqrt{8\pi G\rho}\,v, \quad (140)$$

where the speed of the spacecraft is v.

If there is an additional frequency offset in the Doppler observations from the celestial-mechanics solution it will appear to the fitting program as a slight offset in the orbital parameters. What is observed as an anomalous acceleration is the rate of change of the observed velocity minus the computed acceleration. Since the fitting program does not include the effects of curvature-redshift, it tries to compensate for the apparent velocity offset by a small change in the direction of the trajectory. Since energy is conserved, this velocity offset will produce a small change in effective distance Δr. The energy equation can be written as

$$v^2 = v_\infty^2 + \frac{2\mu}{r}, \quad (141)$$

where v_∞ is the velocity at infinity and $\mu=GM$ is the gravitational constant times the mass of the sun. Then it is apparent that to produce an increase in velocity the computed distance must be less that the true distance. Therefore

$$\Delta v = -\frac{\mu}{r^2 v}\Delta r, \quad (142)$$

and we get

$$v_{observed} - v_{computed} = \alpha r - \frac{\mu}{r^2 v}\Delta r. \qquad (143)$$

To estimate Δr we assume that the trajectory-fitting program has exactly compensated for curvature-redshift at the radius r_0. Thus putting r_0=40 AU where the velocity is v_0=1.251×10^4 m.s^{-1} and then back substituting we get

$$v_{observed} - v_{computed} = \alpha r - \frac{r_0^3 v_0}{r^2 v}\alpha. \qquad (144)$$

Then the apparent acceleration, a, is given by the time derivative and using $\dot{v} = -\mu/r^2$ we get

$$a_0 = -\left[1 - \tfrac{1}{2}\left\{1-\left(v_\infty/v_0\right)^2\right\}\right]v_0\alpha = -0.9096 v_0 \alpha. \qquad (145)$$

Note that as postulated, the celestial mechanics used by the fitting-program has reversed the sign of curvature-redshift to give an acceleration that is in the direction of the sun.

It might appear from equation (140) that when curvature-redshift is expressed as acceleration it is only a function of the dust density at the position of the spacecraft. However, it depends on the integral of the square root of the density between earth and the spacecraft. Furthermore, the batch processing used in fitting the Pioneer 10 parameters ensures that there is a time averaging of the order of 100 days. Thus, we do not expect the results to be sensitive to either spatial or temporal fluctuations in the density.

Using the observed acceleration of 8.74×10^{-10} m.s^{-2}, the required density for the two-way path is 8.8×10^{-19} kg.m^{-3}. The only constituent of the interplanetary medium that approaches this density is dust. One estimate (Le Sergeant & Lamy 1980) of the interplanetary dust density at 1 AU is 1.3×10^{-19} kg.m^{-3} and more recently, Grün (1999) suggests a value of 10^{-19} kg.m^{-3} which is consistent with their earlier estimate of 9.6×10^{-20} kg.m^{-3} (Grün, Zook & Giese 1985). Although the authors do not provide uncertainties it is clear that their densities could be in error by a factor of two or more. The main difficulties are the paucity of information and that the observations do not span the complete range of grain sizes. Also relevant is that the density is required at the distance of Pioneer 10 in 1998 of 72 AU and in the plane of the ecliptic (ecliptic latitude of Pioneer 10 is 3°).

Nieto et al. (2005) have investigated the dust density by seeing if the anomalous acceleration of Pioneer 10 could be due to drag from the dust. Assuming a uniform dust density, they derived an upper limit for the dust density of 3×10^{-16} kg.m^{-3} and lower limits for other density distributions. This density is much higher than the density required here. The meteoroid experiment on board Pioneer 10 measures the flux of grains with masses larger than 10^{-10} g. The results show that after it left the influence of Jupiter the flux (Anderson et al. 1998) was essentially constant (in fact there may be a slight rise) out to a distance of 18 AU. It is thought that most of the grains are being continuously produced in the Kuiper belt. As their orbits evolve inwards due to Poynting-Robertson drag and planetary perturbations, they achieve a roughly constant spatial density. Given the large uncertainties in both the observed density at 1 AU (due to the limitations of the detectors), and the extrapolation of the density to 72 AU, the conclusion is that interplanetary dust could provide the required density to explain the 'anomalous acceleration' by a frequency shift due to curvature-redshift.

Anderson et al. (2005) also find an annual variation with velocity amplitude of $(1.05 \pm 0.01) \times 10^{-4}$ m.s^{-1} and a phase (5.3 ± 7.2) degrees relative to conjunction. Curvature-redshift model predicts a sinusoidal velocity with amplitude $1.23 \times 10^7 \sqrt{\rho_E}$ m.s^{-1} where ρ_E is the density of the medium at the earth's radius. With $\rho_E = 10^{-18}$ kg.m^{-3} the predicted effective velocity is 1.2×10^{-4} m.s^{-1}, a value, which is two orders of magnitude too large. However, the navigation program would interpret the sinusoidal velocity as being due to a pointing error and essentially eliminate it. This will also explain why the phase angle is effectively zero. There is also an indication of a diurnal effect. Since curvature-redshift is approximately proportional to the distance to Pioneer 10 it would predict an effective velocity that is a function of the angle between the position of the receiving ground station and the position of Pioneer 10.

Overall, this analysis has shown that it is possible to explain the acceleration anomaly of Pioneer 10 but that a more definitive result requires curvature-redshift to be included in the fitting program and more accurate estimates of the dust density are certainly needed. Subject to the caveat about the dust density, curvature-redshift can explain the anomaly in the acceleration of Pioneer 10 (and by inference other spacecraft).

9 Conclusion

9.1 Summary of observations

Because of the large number of topics covered it is useful to have a summary of how well curvature-cosmology has fared in explaining the observations. More importantly, what observations may show up problems with curvature-cosmology. Table 24 provides a subjective guide to how well curvature-cosmology can explain the observations. Each topic is scored with a letter where

- A. Quantitative fit to the data is excellent.
- B. Quantitative fit to the data is reasonable with discrepancies mainly due to a poor model for the topic or due to problems with the observations.
- C. Only qualitative agreement.
- D. Neutral as a test – but not likely to invalidate the theory

Table 24: Score card for curvature-cosmology.

Topic	Score	Comments
X-ray background	A	Difficulty with density homogeneity
Cosmic microwave background	B	Good temperature with assumed spectrum
Type 1a supernovae	A	Needs validation of selection effects
Quasar variability	A	Variability is independent of redshift
Size of radio sources	A	Excellent results with small discrepancy due to selection effects
Tolman surface brightness	A	Excellent: much better than Big-Bang
Clusters of galaxies	A	No dark matter: much evidence in support of curvature-redshift
Lyman-alpha forest	D	Major problem with distribution probably due to incorrect line widths
Galaxy distribution	A	Fully consistent without evolution.

Quasar distribution	A	Fully consistent without evolution
Butcher-Oemler effect	B	Is it real?
Radio source counts	A	Basic fit is excellent
Galactic rotation curves	D	Problem with lack of symmetry
Redshifts in our Galaxy.	B	K-effect: scope for more tests
Solar neutrino discrepancy	A	Surprisingly good
Pioneer 10 acceleration	B	Needs accurate dust densities

Probably the most unsatisfying problems are the distribution of the lines in the Lyman-alpha forest and galactic rotation curves. Currently the observations of these topics do not suggest that there is a serious problem with curvature-cosmology but on the other hand, they do not have a complete or satisfactory explanation. Another aspect that needs a full investigation is whether curvature-cosmology can explain the abundances of light elements. Because of the very low density, the abundances may not be in equilibrium and as mentioned, the more important process may be destruction rather than synthesis.

The two major hypotheses of curvature-redshift and curvature-pressure have auxiliary hypotheses of redshift inhibition, curvature-redshift from particles other than photons and local curvature-pressure. Curvature-redshift and curvature-pressure are intrinsic to curvature-cosmology and are involved in explaining nearly all the observations. Thus, the support of curvature-cosmology by the observations is also support for these hypotheses. The auxiliary hypothesis of curvature-redshift-inhibition is mainly supported by the non-observance of curvature-redshift in laboratory experiments. The other field where the inhibition is important is in the observation of redshifts within our own Galaxy. At present, the main evidence is negative in that if curvature-redshift-inhibition was invalid, the model of the Galactic structure derived from radio frequency surveys would be much more complex. Positive support would come from different redshifts between radio and optical spectral lines for the same

Summary of observations 149

distant object. The application of curvature-redshift to electrons is tested by the prediction of the cosmic microwave background radiation. The idea of local curvature-pressure is central to the analysis for solar neutrino production. Thus, the agreement of the theory with these observations can be taken as support for both the primary and the secondary hypotheses.

As well as the redshifts already discussed there many reports of other non-Doppler redshifts. Arp (1987) has reported on many cases where there is a apparent physical connection between a galaxy and a quasar that have significantly different redshifts. Needless to say, these connections are disputed and even if they are correct, curvature-cosmology offers no resolution. It is hoped that there are more studies of these associations without the acrimonious debate that has already occurred. Reboul (1981) provides list of some 780 publications on "untrivial redshifts". Whether curvature-redshift can offer any explanation for these redshifts depends on whether they are confirmed and of knowledge of the local gas densities.

Overall the agreement with observations is excellent and in nearly every case much better that that achieved by Big-Bang cosmology. More tellingly, the comparison is done with none or very few free parameters. However, I suspect that the major objection to curvature-cosmology will be theoretical. Although I believe that it is in full agreement with general relativity and quantum theory curvature-cosmology does require new concepts. I can only hope that this work will stimulate further work that will lead to this or a better theory being used to increase our understanding of the universe.

References

Abramowitz, M. and Stegun, I. A. (1972) 'Handbook of Mathematical Functions' (Dover: New York)
Allen, C. W. (1976) 'Astrophysical Quantities', 3rd Ed (Athlone: London)
Anderson, J.D, Laing, P.A., Lau, E.L., Liu, A.S., Nieto, M.M., Turyshev, S.G. (1998) Phys. Rev. Let., **81**, 2858
Anderson, J.D., Lau, E.L., Scherer, K., Rosenbaum, D.C., Tepliz, V.L. (1998) Icarus, **131**, 167
Anderson, J.D, Laing, P.A., Lau, E.L., Liu, A.S., Nieto, M.M., Turyshev, S.G. (2002) Phys. Rev., **D65**, 82004
Andreon, A., Lobo, C., Iovino, A. (2004) Mon. Not. R. Astr. Soc., **349**, 889
Arp, H. (1987) 'Quasars, Redshifts and Controversies' (Interstellar Media: Berkeley, California)
Arp, H. (1992) Mon. Not. R. Astr. Soc., **258**, 800
Aschwanden, M.J. (2004) 'Physics of the Solar Corona' (Springer: New York)
Astier, A., Guy, J., Regnault, N., Pain, R., Aubourg, E., Balam, D., Basa, S., Carlberg, R.G., Fàbbro, S., Fouchez, D., Hook, I.M., Howell, D.A., Lafoux, H., Neill, J.D., Palanque-Delabrouille, N., Perŕet, K., Pritchet, C.J., Rich, J., Sullivan, M., Taillet, R., Aldering, G., Antilogus, P., Arsenijevic, V., Balland, C., Baumont, S., Bronder, J., Courtois, H., Ellis, R.S., Filiol, M., Gonçalves, A.C., Goobar, A., Guide, D., Hardin, D., Lusset, V., Lidman, C., McMahon, R., Mouchet, M., Mourao, A., Perlmutter, S., Ripoche, P., Tao, C., Walton, N. (2005) Astr. Astrophys., **207**, 1504
Bahcall, J.N. (1989) 'Neutrino Astrophysics' (Cambridge University Press)
Bahcall, J.N., Pinsonneault, M.H., Basu, S. (2001) Astrophys. J., **555**, 990
Bahcall, J.N., Serenilli, A. M., Pinsonneault M. (2004) Astrophys. J., **614**, 464
Battistelli, E.S., De Peter, M., Lamagna, L., Melchiorre, F., Palladino, E., Savini, G, Cooray, A., Melchiorre, A., Rephaeli, Y., Shimin,M. (2002) Astrophy. J. **580**, L104
Beijersbergen, M. (& van der Hulst, J.M.) (2004) Phd Thesis, University of Groningen
Bell, E.F., Wolf, C., Meisenheimer, K., Rix, H.-W., Borch, A., Dye, S., Kleinheinrich, M., Wisotzki, L., McIntosh, D.H. (2004) Astrophys. J. **608**, 752
Bennett, C.L., Lawrence, C.R., Hewitt, J.N., Burke, B.F. (1983) BAAS., **15**, 935
Bondi, M., Ciliegi, P., Zamorani, G., Gregorini, L., Vettolani, G., Parma, P., de Ruiter, H., Le Fevre, O., Arnaboldi, M., Guzzo, L., Maccagni, D., Scaramella, R., Adami, C., Bardelli, S., Bolzonella, M., Bottini, D., Cappi, A., Foucaud, S., Franzetti, P., Garilla, B., Gwyn, S., Ilbert, O., Iovino, A., Le Brun, V., Marano, B., Marinoni, C., McCracken, H.J., Menaux, B., Pollo, A., Pozzetti, L., Radovich, M., Ripepi, V., Rizzo, D., Scodeggio, M., Tresse, L., Zanichelli, A., Zucca, E. (2003) Astr., Astrophys., **403**, 857
Born, M., Wolf, E. (1999) 'Principles of Optics' (Cambridge University Press)

Braginskiĭ, V.B., Panov, V.I. (1971) Zh. Eksp. Teo. Fiz. **61**, 873 [Sov. Phys. JETP **34**, 463 (1972)]
Briel, U.G., Henry, J.P, Böhringer, H. (1992) Astr., Astrophys., **259**, L31
Buchalter, A., Helfand, D.J., Becker, R.H., White, R.L. (1998) Astrophy. J., **494**, 503
Butcher, H., Oemler, A.Jr. (1978) Astrophys. J., **219**, 18
Butcher, H., Oemler, A. Jr. (1984) Astrophys. J., **285**, 426
Cassinelli, J.P. (1979), Ann. Rev. Astr. Astrophys., **17**, 275
Colla, G., Fanti, C., Fanti, R., Gioia, C., Lari, C., Lequeux, J., Lucas, R., Ulrich, M.H. (1975) Astron. Astr. Suppl. **20**, 1
Colless, M., Dunn, A.M. (1996) Astrophys. J., **458**, 435
Cowsik, R., Kobetich, E.J. (1972) Astrophys. J., **177**, 585
Crawford, D.F. (1975) Nature, **254**, 313
Crawford, D.F. (1979) Nature, **277**, 633
Crawford, D. F. (1987a) Aust. J. Phys., **40**, 440
Crawford, D. F. (1987b) Aust. J. Phys., **40**, 459
Crawford, D. F. (1991) Astrophys. J., **377**, 1
Crawford, D. F. (1993) Astrophys. J., **410**, 488
Crawford, D. F. (1995a) Astrophys. J. **440**, 466
Crawford, D. F. (1995b) Astrophys. J., **441**, 488
Crawford, D. F. (1995) Aust. J. Phys., **52**, 753
Cristiani, S., Vio, R. (1990) Astr. Astrophys., **227**, 385
Croom, S.M., Smith, R.J., Boyle, B.J., Shanks, T., Miller, L., Outram, P.J., Loaring, N.S. (2004) Mon. Not. R. Astr. Soc. **349**, 1397
de Groot, S.R., Leeuwen, W.A., van Weert, C.G. (1980) 'Relativistic Kinetic Theory' (Amsterdam: North-Holland)
Dennis, B.R., Suri, A.N., Frost, K.J. (1973) Astrophys. J., **186**. 97
De Propris, R., Colless, M., Driver, S.P., Couch, W., Peacock, J.A., Baldry, I.K., Baugh, C.M., Bland-Hawthorn, J., Bridges, T., Cannon, R., Cole, S., Collins, C., Cross, N., Dalton, G.B., Efstathiou, G., Ellis, R.S., Frenk, C.S., Glazebrook, K., Hawkins, E., Jackson, C. Lahav, O., Lewis, I., Lumsden, S., Maddox, S., Madgwick, D.S., Norberg, P., Percival, W., Peterson, B., Sutherland, W., Tayloer, K. (2003) Mon. Not. R. Astr. Soc., **242**, 725
De Propris, R., Stanford, A.A., Eisenhardt, P.R., Dickinson, M. (2003) Astrophys. J., **598**, 20
Dicke, R.H. (1964) Nature, **202**, 432
D'Inverno, R. (1992). 'Introducing Einstein's Relativity' (Clarendon Press: Oxford)
Djorgovski, S.G., Spinrad, H. (1981) Astrophys. J. **251**, 417
Ellis, G.G.R. (1984) Ann. Rev. Astr. Astrophys., **22**, 157
Eötvös, R.V., Pekar, D., Fekete, E. (1922) Ann. Phys. (Leipzig) **68**, 11
Faranoff, B.L., Riley, J.M. (1974) Mon. Not. R. Astr. Soc., **167**, 31
Feynman, R.P., Leighton, R.B., Sands, M. (1965) 'The Feynman Lectures on Physics' (Addison-Wesley: Massachusetts)
Field, G.B., Henry, R.C. (1964) Astrophys. J., **140**, 1002
Finoguenov, A., Briel, U.G., Henry, J.P., Gavazzi, G., Iglesias-Paramo.J, Boselli, A. (2004) Astr., Astrophys., **419**, 47
Fixsen, D.J., Cheng, E.S, Gales, J.M., Mather, J.C., Shafer, R.A., Wright, E.L. (1996) Astrophys. J., **473**, 576

Index

Fomalont, E. B., Kellermann, K. I., Wall, J. V., Weistrop, D. (1984) Science, **225**, 23

Freedman, W.L., Madore, B.F., Gibson, B.K., Ferrarese, L., Kelson, D.D., Sakai, S., Mould, J.R., Kennicutt, R.C. Jr., Ford, H.C., Graham, J.A., Huchra, J.P., Hughes, S.M.G., Illingworth, G.D., Macri, L.M., Stetson, P.B. (2001) Astrophys. J., **553**, 47

Fukada, Y., Hayakawa, S., Ikeda, M., Kasahara, J., Makino, F., Tanaka, Y. (1975) Astrophys. Space Sci., **32**, L1

Fukugita, M., Ichikawa, T., Gunn, J.E., Doi, M., Shimasaku, K., Schneider, D.P. (1996) Astr. J. **111**, 1748

Ge, J., Bechtold, J., Black, J.H. (1995) Astrophys. J. **474**, 67

Ghosh, A. (1991) Aperion **9-10**, 95

Giacconi, R., Gursky, H., Paolini, F.R., Rossi, B.B. (1962) Phys. Rev. Letters, **9**, 439

Goldhaber, G., Groom, D.E., Kim, A., Algering, G., Astier, P., Couley, A., Deustua, S.E., Ellis, R., Fabbro, S., Fruchter, A.S., Goobar, A., Hook, I., Orwin, M., Kim, M., Knop, R.A., Lidman, C., McMahon, R., Nugent, P.E., Pain, R., Panagia, N., Pennypacker, C.R., Perlmutter, S., Ruiz-Lapuente, P., Schaefer, B., Walton, N.A., York, T. (2001) Astrophys. J. **558**, 359

Goldstein, H. (1980) 'Classical Mechanics' (Addison-Wesley: Massachusetts)

Goobar, A., Perlmutter, S. (1995) Astrophys. J., **450**, 14

Gould, R.J., Burbidge, G.R. (1963) Astrophys. J., **138**, 969

Gruber, D.E., Matteson, J.L., Peterson, L.R. (1999) Astrophys. J., **520**, 124

Grün, E., Zook, H.A., Giese, R.H. (1985) Icarus **62**, 244

Grün, E. (1999) in 'Encyclopaedia of the Solar System', edited by P.R. Weismann, L.-A. McFadden, and T. V. Johnston (Academic Press) 777

Gruppioni, C., Cilieg1, P., Rowan-Robinson, M., Cram, L., Hopkins, A., Cesarsky, C., Danese, L., Franceschini, A., Genzel, R., Lawrence, A., Lemke, D., McMahone, R.G., Miley, G., Oliver, S., Puget, j.-L., Rocca-Volmerange, B. (1999) Mon. Not. R. Astr. Soc., **305**, 297.

Gurvits, L.I., Kellermann, K.I., Frey, S. (1999) Astr. Astrophys., **342**, 378

Guth, J.E. (`1981) Phys. Rev. D, **23**, 347

Hamuy, M., Phillips, M.M., Schommer, R.A., Suntzeff, N.B. (1996). Astr. J., **112**, 2391

Havas, P. (1966) Am. J. Phys., **34**, 753

Hawkins, M.R.S. (2001) Astrophys. J., **553**, L97

Hawkins, M.R.S. (2003) Mon. Not. R. Astr. Soc., **344**, 492

Hendry, M.A., Simmonds, J.F.L., Newsam, A.M. (1993) 'Cosmic Velocity Fields', Proceedings of the 9th IAP Astrophysics Meeting, Institut d'Astrophysique, Paris, July 12-17, 1993. Edited by F. R. Bouchet and M. Lachièze-Rey. Gif-sur-Yvette: Editions Frontieres, p.23, also astro-ph/9310028

Hillebrandt, W., Niemer, J.C. (2000) Ann. Rev. Astr. Astrophys., **38**, 191

Hogg, D.W. (1999) astro-ph/9905116

Hogg, D.W., Baldry, I.K., Blanton,M.R., Eisensteint, D.J. (2002) astr-ph/0210394

Holt, S.S. (1992) in 'The X-ray Background', ed. X. Barcons, A. C. Fabian (Cambridge University Press) 29

Index

Hopkins, A.M., Mobasher, B., Cram, L., Rowan-Robinson, M. (1998) Mon. Not. R. Astr. Soc., **296**, 839
Horstman-Morretti, E., Fuligni, F., Horstman, H.M., Brini, D. (1974) Astrophys. Space Sci., **27**, 195
Hoyle, F., Narlikar, J.V. (1963) Astrophys. J., **137**, 993
Hughes, J.P. (1989) Astrophys. J., **337**, 21
Itoh, N., Sakamoto, T., Kusano, S., Nozawa, S., & Kohyama, Y. (2000) Astrophys. J. Suppl., **128**, 125
Jackson, J.D. (1975) 'Classical Electrodynamics' (John Wiley & Sons: New York)
Jauch J.M. and Rohrlich F. 1980. 'The Theory of Photons and Electrons', 2nd edition, (Springer Verlag: New York)
Karoji, K., Nottale, L., Vigier, J-P. (1976) Astrophys., Space Sci., **44**, 229
Kellerman, K.I., Wall, J.V. (1987) in 'Observational Cosmology' ed. A. Hewitt et al. (D. Reidel Publishing Co: Dordrecht) (IAU Symposium, No. 124) 545
Kibblewhite, E. J., Bridgeland, M. T., Bunclark, P. S., Irwin, M. J. (1984) 'Astronomical Microdensitometry Conference' NASA CP-2317, ed. D.A. Klinglesmith (NASA: Washington) p277
Kim, A., Goobar, A., Perlmutter, S. (1996) Proc. Astr. Soc. Pacif., **108**, 190
Kinzer, R.L., Jung, G.V., Gruber, D.E., Matterson, J.L., Peterson, L.E. (1997) Astrophys. J., **475**, 361
Knop, R.A., Aldering, G., Amanullah, R., Astier, P., Blanc, G., Burns, M.S., Conley, A., Deustua, S.E., Doi, M., Ellis, R., Fabbro, S., Folatelli, G., Fruchter, A.S., Garavini, G., Garmond, S., Garton, K., Gibbons, R., Goldhaber, G., Goobar, A., Groom, D.E., Hardin, D., Hook, I., Howell, D.A., Kim, A.G., Lee, B.C., Lidman, C., Mendez, J., Nobili, S., Nugent, P.E., Pain, R., Panagia, N., Pennypacker, C.R., Perlmutter, S., Quimby, R., Raux, J., Regnault, N., Ruiz-Lapuente, P., Sainton, G., Schaefer, B., Schahmabeche, K., Smith, E., Spadafora, A.L., Stanishev, V., Sullivan, M., Walton, N.A., Wang, L., Wood-Vasey, W.M., Yasuda, N. (2003) Astrophys. J., **598**, 102
Langston, G.I., Heflin, M.B., Conner, S.R., Lehár, J., Carrili, C.L., Burke, B.F. (1990) Astrophys. J. Suppl. Ser., **72**, 621
Lapparent, V., Geller, M.J., Huchra, J.P. (1986) Astrophys. J., **302**, L1
LaViolette, P.A. (1986) Astrophys. J. **301**, 544
Le Sergeant D'Hendecourt, L.B. Lamy, Ph.L. (1980). Icar. **43**, 350
Liddle, A.E., Lyth, D.H. (2000) 'Cosmological Inflation and Large-Scale Structure' (Cambridge University Press)
Lima, J.A.S, Silva, A.J., Viegas, S.M. (2000) Mon. Not. R. Astr. Soc. **312**, 747
Liu, M.C., Graham, J.R. (2001) Astrophys. J., **557**, L31
Longair, M.S. (1991). 'Theoretical Concepts in Physics' (Cambridge University Press)
Lubin, L.M., Sandage, A. (2001a) Astr. J., **121**, 2289
Lubin, L.M., Sandage, A. (2001b) Astr. J., **122**, 1071
Lubin, L.M., Sandage, A. (2001c) Astr. J., **122**, 1084
Malmquist, K. (1920) Medd. Lund. Astr. Bs., **22**, 1
McDonald, A.B. (2004) New J. Phys., **6**, 121
McGilchrist, M.M., Baldwin, J.E., Riley, J.M. Titterington, D.J., Waldram, E.M., Warner, P.J. (1990) Mon. Not. R. Astr. Soc., **246**, 110

Index

Mandrou, P., Vedrenne, G., Niel, M. (1979) Astrophys. J., **230**, 97
Marmet, P. (1988) Physics Essays, **1**, 24
Marmet, P. & Reber, G. (1989) ITPS, **18**, 264
Marshall, F.E., Boldt, S.S., Miller, R.B., Mushotzky, R.F., Rose, L.A., Rothschild, R.E., Sermlemitios, P.J.1980) Astroph. J., **235**, 4
Maslowski, J., Pauliny-Toth, I.I.K., Witzel, A., Kühr, H. (1984a) Astr., Astrophys., **139**, 85
Maslowski, J., Pauliny-Toth, I.I.K., Witzel, A., Kühr, H. (1984b) Astr., Astrophys., **141**, 376
Mather, J.C., Cheng, E.S., Epler, Jr., R.E., Isaacmsn, R.B., Meyer, S.S., Shafer, R.A., Wiess, R., Wright, E.L., Bennett, C.L., Boggess, N.W., Dwek, E., Gulkis, S., Hauser, M.G., Janssen, M., Kelsall, T., Lubin, P.M., Moseley, S.H., Murdock, T.L., Silverberg, R.F., Smoot, G.F., Wilkinson, D.T. (1990) Astrophys. J., **354**, L37
Mather, J.C., Fixsen, D.J., Shafer, R.A., Mosier, C., Wilkinsob, D.T. (1999) Astrophys. J., **512**, 511
Mayall, N.U. (1960) Annales d'Astrophysique, **23**, 344
Mazets, E.P., Golenetskii, S.V., Il'inskii, V.N., Gur'yan, Yu.A., Kharitonova, T.V. (1975) Astrophys. Space Sci., **33**, 347
Misner, C.W., Thorne, K.S., and Wheeler, J.A. (1973). 'Gravitation' (W. H. Freeman: San Francisco)
Nieto, M.N., Turyshev, S.G., Anderson, J.D. (2005) astro-ph/051626
Nottale, L. (1976) Astrophys. J., **208**, L103
Nozawa, S., Itoh, N., Kohyama, Y. (1998) Astrophys. J., **507**, 530
Parijskij, Yu.N., Bursov, N.N., Lipovka, N.M., Soboleva, N.S., Temirova, A.V. (1991) Astr. Astrophys. Suppl. Ser., **87**, 1
Partridge, R.B., Hilldrup, K.C., Ratner, M.I. (1986) Astrophys. J., **308**, 46
Pearson, T.J., Kus, A.J. (1978) Mon. Not. R. Astr. Soc., **182**, 273.
Pecker, J-C,& Vigier, J-P (1987) IAUS, **124**, 507.
Peebles, P.J.E. (1993) 'Principles of Physical Cosmology' (Princeton: New Jersey)
Perlmutter, S., Gabi, S., Goldhaber, G., Goobar, A., Groom, D.E., Hook, I.M., Kim, A.G., Kim, M.Y., Lee, J.C., Pain, R., Pennypacker, C.R., Small, I.A., Ellis, R.S., McMahon, R.G., Boyle, B.J., Bunclark, P.S., Carter, D., Irwin, M.J., Glazebrook, K., Newberg, H.J.M, Filippenko, A.V.,Matheson, T., Dopita, M., Couch, W.J. (1997) Astrophys. J., **483**, 565
Perlmutter, S., Aldering, G., Goldhaber, G., Knop, R.A., Nugent, P., Castro, P.G., Deustua, S., Fabbro, S., Goobar A., Groom, D.E., Hook, I.M., Kim, A.G., Kim, M,Y., Lee, J.C, Nunes, N.J., Pain, R., Pennypacker, C.R., Quimby, R., Lidman, C., Ellis, R.S., Irwin, M., McMahon, R.G., Ruiz-Lapuente, P., Walton, N., Shaefer, B., Boyle, B.J., Filippenko, A.V., Matheson, T., Fruchter, A.S., Panagia, N., Newberg, H.J.M., Couch, W.J. (1999) Astrophys. J., **517**, 565
Perlmutter, S., Schmidt, B.P. (2003) in 'Supernovae, Gamma Ray Bursts', K. Weiler, ed. (Springer: New York)
Petrosian, V. (1976) Astrophys. J., **209**, 1
Pettini, M., Hunstead, R.W., Smith, L.J., Mar, D.P. (1990) Mon. Not. R. Astr. Soc., **246**, 545
Phillips, M.M. (1993) Astrophys. J., **413**, L105

Pimbblet, K.A. (2003) Pub. Astr. Soc. Aust., **20**, 294
Postman, M., Lauer, T.R. (1995) Astrophys. J., **440**, 28
Plionis, M., Tovmassian, H.M. (2004) Astr., Astrophys., **416**, 441
Prandoni, I., Gregorini, L., Parma, P., de Ruiter, H.R., Vettolani, G., Wieringa, M.H., Ekers, R.D. (2001) Astr., Astrophys., **365**, 392
Pound, R.V., Snyder, J.L. (1965) Phys. Rev. B, **140**, 788
Press, W.H., Teukolsky, A.A., Vetterling, W.T., Flannery, B.P. (1992) 'Numerical Recipes in Fortran 77' (Cambridge University Press)
Rauch, M. (1998) Ann. Rev. Astr. Astrophys., **36**, 267
Raychaudhuri, A.K. (1955) Phys. Rev., **98**, 1123
Reboul, H.J. (1981) Astr. Astrophys., **45**, 129
Redfield, S. (2006) astro-ph/0601117
Richardson, D., Branch, D., Casebeer, D., Millard, J., Thomas, R. C. & Baron, E. (2002) Astron. J. **123**, 745
Riess, A.G., Strolger, L-G., Tonry, J., Casertano, S., Fergusen, H.C., Mobasher, B., Challis, P., Filippenko, A.V., Jha, S., Li, W., Chornock, R., Kirshner, R.P., Leibundgut, B., Dickinson, M., Livio, M., Giavalisco, M., Steidel, C.C., Benitez, N., Tsvetanov, Z. (2004) Astrophys. J., **607**, 645
Riess, A.G., Li, W., Stetson, P.B., Filippenko, A.V., Jha, S., Kirshner, R.P., Challis, P.M., Garnavitch, P.M., Chornock, R. (2005) Astrophys. J., **627**, 579
Rindler, W. (1977) 'Essential Relativity' (Springer-Verlag: New York)
Robertson, J.G. (1973) Aust. J. Phys., **26**, 403
Robertson, J.G. (1977a) Aust. J. Phys., **30**, 209
Robertson, J.G. (1977b) Aust. J. Phys., **30**, 231
Rood, H.J. (1988) Ann. Rev. Astr. Astrophys., **26**, 245
Roth, K.C., Meyer, D.M. (1995) Astrophys. J. **441**, 129
Rowan-Robertson, M (1985) 'The Cosmological Distance Ladder' (W.H. Freeman & Co.: New York)
Russel, D.G (2002) Astrophys. J., **565**, 681
Sandage, A., Lubin, L.M. (2001) Astr. J., **121**, 2271
Sandage, A., Perelmuter, J-M. (1990) Astrophys. J., **370**, 455
Schmidt, M. (1968) Astrophys. J. **151**, 393
Schatzman, E (1979) Astron. & Astrophys., **74**, 12
Schneider, D.P., Fan, X., Hall, P.B., Jester, S., Richards, G.T., Stoughton, C., Strauss, M.A., SubbaRao, M., Vanden Berk, D.E., Anderson, S.F., Brandt, W.N., Gunn, J.E., Gray, J., Trump, J.R., Voges, W., Yanny, B., Bahcall, N.A., Blanton, M.R., Boroski, W.N., Brinkmann, J., Brunner, R., Burles, S., Castander, F.J., Doi, M., Eisenstein, D., Frieman, J.A., Fugugita, M., Hecjman, T.M., Hennesy, G.S., Iveskić, Z., Kent, S., Knapp, G.R., Lamb, D.Q., Lee, B.C., Loveday, J., Lupton, R.H., Margon, B., Meiksin, A., Munn, J.A., Newberg, H.J., Nichol, R.C., Niederste-Ostholt, M., Pier, J.R., Richmond, M.W., Rockosi, C.M., Saxe, D.H., Schlegel, D.J., Szalay, A.S., Thaker, A.R., Uomoto, A., York, D.G. (2003) Astr. J., **126**, 2579
Schneider, D.P., Gunn, J.E, & Hoessel, J.G. (1983) Astrophys. J. 264, 337.
Schneider, D.P., Hall, P.B., Richards, G.T., Vanden Berk, D.E., Anderson, S.F., Fan, X., Jester, S., Stoughton, C., Strauss, M.A., SubbaRao, M., Brandt, W.N., Gunn, J.E., Yanny, B., Bahcall, N.A., Barentine, J.C., Blanton, M.R., Boroski, W.N., Brewington, H.J, Brinkmann, J., Brunner, R., Csabai, I.,

Doi, M., Eisenstein, D.J., Friemann, J.A., Fukugita, M., Gray, J., Harvanek, M., Heckman, T.M., Iveskić, Ž., Kent, S., Kleinman, S.J., Knapp, G.R., Kron, R.G., Krzesinski, J., Long, D.C., Loveday, J., Lupton, R.H., Margon, B., Munn, J.A., Neilsen, E.H., Newberg, H.J., Newman, P.R., Nichol, R.C., Nitta, A., Pier, J.R., Rockosi, C.M., Saxe, D.H., Schlegel, D.J., Snedden, S.A., Szalay, A.S., Thaker, A.R., Uomoto, A., York, D.G. (2005) Astr. J. **130**, 367

Sciama, D.W. (1953) Mon. Not. R. Astr. Soc., **113**, 34

Sciama, D.W. (1971) 'Modern Cosmology", Cambridge University Press)

Srianand, R., Petitjean, P. & Ledoux, C. (2000) Nature **408**, 931

Steinhardt, P.J., Caldwell, R.R (1998) 'Cosmic Microwave Background & Large Scale Structure of the Universe', Astr. Soc. Pacif. Conf. Ser. **151**, 13

Strauss, M. A., Gunn, J.E. (2001) http://www-wfau.roe.ac.uk/sdss/documents/response.dat

Strolger, L.-G., Riess, A.G., Dahlen, T., Livio, M., Panagia, N., Challis, P., Tonry, J.L., Filippenko, A.V., Chornock, R., Fergusen, H., Koekemoer, A., Mobasher, B., Dickinson, M., Giavalisco, M., Casertano, S., Hook, R., Blondin, S., Leibundgut, B., Nonino, M., Rosati, P., Spinrad, H., Steidel, C.C., Stern, D., Garnavich, P.M., Matheson, T., Grogin, N., Hornschemeir, A., Kretchmer, C., Laidler, V.G., Lee, K., Lucas, R., de Mello, D., Moustakas, L.A., Ravindranath, S., Richardson, M., Taylor, E. (2004) Astrophys. J., **613**, 200

Subrahmanya, C.R., Mills, B.Y. (1987) 'Observational Cosmology' ed. A. Hewitt, G. Burbidge and Li Zhi Fang. (D. Reidel Publishing Co.: Dordrecht) (IAU Symposium, No. 124) 569

Sunyaev, R.A., Zel'dovich, Ya.B. (1970) Astrophys. Space Sci. **7** 3

Taylor, A.C., Grainge, K., Jones, M.E., Pooley, G.G., Saunders, D.E., Waldram, E.M. (2001) Mon. Not. R. Astr. Soc., **327**, L1

Tolman, R.C. (1934) 'Relativity, Thermodynamics and Cosmology' (Oxford University Press)

Trombka, J.L., Dyer, C.S., Evans, L.G., Bielefeld, M.J., Seltser, S.M., Metzger, A.E. (1977) Astrophys. J., **212**, 925

Turner, M.S. (1999) The Third Stromlo Symposium: 'The Galactic Halo', Gibson, B.K., Axelrod, T.S., Putman, M.E. (eds) ASTR. SOC. PACIF. Conf. Ser. **165**, 431

Vanden Berk, D.E., Richards, G.T., Bauer, A., Strauss, M.A., Schneider, D.P., Heckman, T.M., York, D.G., Hall, P.B., Fan, X., Knapp, G.R., Anderson, S.F., Annis, J., Bahcall, N.A., Bernardi, M., Briggs, J.W., Brinkmann, J., Brunner, R., Burles, S., Carey, L., Castander, F.J., Connolly, A.J., Crocker, J.H., Csabal, I., Doi, M., Finkbeiner, D., Friedman, S., Frieman, J.A., Fugugita, M., Gunn, J.E., Hennessy, G.S., Ivezić, Ž., Kent, S., Kunszt, P.Z., Lamb, D.Q., Leger, R.F., Long, D.C., Loveday, J., Lupton, R.H., Meiksin, A., Merelli, A., Munn, J.A., Newberg, H.J., Newcomb, M., Nichol,R.C., Owen, R., Pier, J.R., Pope, A., Rockosi, C.M., Schlegel, DL., Siegmund, W.A., Smee, S., Snir, Y., Stoughton, C., Stubbs, C., Subba Rao, M., Szalay, A.S., Szokoly, G,P., Tremonti, C., Uomotto, A., Waddell, P., Yanny, B., Zheng, Wei (2001) Astr. J., **122**, 549

Vanden Berk, D.E., Wilhite, B.C., Kron, R.G., Anderson, S.F., Brunner, R.J., Hall, P.B., Izevezić, Ž., Richards, G.T., Schneider, D.P., York, D.G.,

Brinkmann, J.V., Lamb, D.Q., Nichol, R.C., Schlegel, D.J. (2004) Astrophys. J. **601**, 692
Wall, J.V., Peacock, J.A. (1985) Mon. Not. R. Astr. Soc., **216**, 173
Watt, M.P., Ponman, T.J., Bertram, D., Eyles, C.J., Skinner, G.K., Willmore, A.P. (1992) Mon. Not. R. Astr. Soc., **258**, 738
Weinberg, S. (1972) 'Gravitation and Cosmology' (Wiley: New York)
White, R.L., Becker, R.H., Helfand, D.J., Gregg, M.G. (1997) Astrophys. J., **475**, 479
White, S.D.M., Briel, U.G., Henry, J.P. (1993) Mon. Not. R. Astr. Soc., **261**, L8
Windhorst, R.A., Fomalont, E.B., Partridge, R.B., Lowenthal, J.D. (1993) Astrophys. J., **405**, 498
Wisotzki, L. (2000) Astr., Astrophys., **353**, 861
Wolf, C., Meisenheimer, K., Rix, H. -W., Borch, A., Dye, S., Kleinheinrich, M. (2003) Astr. Astrophys., **401**, 73
Wolf, C., Meidenheimer, K., Kleinheinrich, M., Borch, A., Dye, S., Gray, M., Wisotski, L., Bell, E.F., Rix, H.-W., Cimatti, A., Hasinger, G., Szokoly, G. (2004) Astr. Astrophys., **421**, 913
Wu, X-P., Xue, Y-J. (1999) Astropys. J., **524**, 22
Zel'dovich, Ya. B. (1963). *Usp. Fiz. Nauk.*, **80**, 357;(1964) Sov. Phys.-Uspekhi. **6**, 475

Index

accretion disks 10
alternative theories 10
angular size 49
astrophysical jets 54
Big Bang
    parameters 47
black hole 38
black holes 9, 54
bremsstrahlung
    background X-rays 62
Butcher-Oemler effect 129
Calàn/Tololo
    supernovae 77
cluster of galaxies
    conclusion 108
clusters of galaxies 100
    virial theorem 9, 100
CMBR
    cosmic microwave background
        radiation 69
    predicted temperature 73
CMBR temperature
    at large redshifts 73
Coma cluster 101
    velocity dispersion 102
cosmic microwave background
    radiation 69
cosmic plasma
    density, temperature 69
cosmological model
    simplest 15
critical density 9
curvature-cosmology 2
    area, volume 43
    geometry 43
    model 39
    scorecard 147
    stability 41
curvature-pressure 2, 14
    derivation 31
    General relativistic model 34
    local 36
    Newtonian model 32
curvature-redshift 2, 12
    derivation 17

electrons and other particles ... 28
    inhibition 25
    secondary photons 23
dark energy 9
dark matter 9
distance modulus 47
entropy 50
extinction length 26
flatness problem 9
focussing theorem 12, 21
Friedmann equations 40
galactic cluster
    intra-cluster gas 106
    X-ray luminosity 105
galactic cluster velocites
    excellent prediction 102
galactic clusters
    seen through
        excess velocity 107
    voids before 107
galactic rotation 136
galaxy distributions 114
galaxy survey
    COMBO-17 114
geodesics deviation
    equation for 18
gravitation force
    misconception 31
horizon problem 9
Hubble constant
    from X-ray background 69
    observations 108
    reduced 47
    theory 42
inflation 9
instability
    of static Einstein model 41
K-corrections
    supernovae 77
K-term
    stellar 138
laboratory tests 27
large number coincidences 55
linear size
    radio sources 92

Index

Lyman alpha forest
    extra velocity in gas 111
    line widths 112
Lyman-alpha forest 110
Mach's principle 56
Malmquist bias 59
Milky Way
    redshifts 137
nuclear abundances 53
Olber's Paradox 52
perfect cosmological principle 16
Petrosian metric radii 96
photon
    single 19
Pioneer 10
    anomalous acceleration 142
    diurnal variation 145
plasma frequency 24
Pound & Snyder
    Gamma ray experiment 27
principle of equivalence 32
quasar
    colour differences 121
    K corrections 122
    reference spectrum 121
quasar catalogue
    SDSS 119
quasar distribution 119
quasar evolution 127
quasar survey
    2QZ+6QZ 120
    SDSS 121
quasars
    SDSS
        scatter plot 124
        variability in magnitude 91
radio source counts 130
radio sources
    intrinsic distribution 132
    linear size 92
redshift
    Hubble-Big Bang 1
    unreliable nearby 61
References 150
regression
    errors in both variable 58
Ricci tensor 21
Riemann-Christoffel tensor 21
Robertson-Walker metric 40
secondary photons
    energy distribution 24
selection effects
    Malmquist bias 59
Sloan Digital Sky Survey
    SDSS 119
solar corona
    heating 141
solar neutrino production 139
spectral lines
    possible bias 61
spiral galaxies
    rotation curves 9
stretch factors
    supernovae 78
Sunyaev-Zel'dovich effect ... 62, 74
Supernova Cosmology Project
    SCP 77
supernovae 75
    theoretical model 80
supernovae analysis
    summary 89
Supernovae Legacy Survey
    SNLS 88
supernovae magnitudes 86
surface brightness 95
    conclusion 99
tired-light
    objections 13
tired-light theories 10
Tolman surface brightness 95
universal plasma
    temperature 41
universe
    stability 14
Virgo cluster 104
wave packet 19, 21
X–ray background 61

www.ingramcontent.com/pod-product-compliance
Lightning Source LLC
Chambersburg PA
CBHW031055180526
45163CB00002BA/848